James Irvine Lupton

The Illustrated Horse Management

James Irvine Lupton

The Illustrated Horse Management

ISBN/EAN: 9783742833747

Manufactured in Europe, USA, Canada, Australia, Japa

Cover: Foto ©berggeist007 / pixelio.de

Manufactured and distributed by brebook publishing software
(www.brebook.com)

James Irvine Lupton

The Illustrated Horse Management

MAYHEW'S
ILLUSTRATED HORSE MANAGEMENT:

CONTAINING DESCRIPTIVE REMARKS UPON

ANATOMY; MEDICINE; SHOEING; TEETH
FOOD; VICES; STABLES;

LIKEWISE A PLAIN ACCOUNT OF

The Situation, Nature, and Value of the various Points;

TOGETHER WITH COMMENTS ON

GROOMS — BREEDERS — BREAKERS AND TRAINERS.

EMBELLISHED WITH NUMEROUS ENGRAVINGS.

REVISED AND IMPROVED BY

JAMES IRVINE LUPTON, M.R.C.V.S.

AUTHOR OF SEVERAL WORKS ON VETERINARY SCIENCE AND ART.

LONDON:
WM. H. ALLEN & CO., 13, WATERLOO PLACE,
PALL MALL, S.W.
1876.

LONDON:

WYMAN AND SONS, PRINTERS, GREAT QUEEN STREET,

LINCOLN'S INN FIELDS, W.C.

PREFACE.

THE statements made in the following chapters become self-evident to the man who can release his mind from the trammels of conventionality, and will allow his practice to be based and guided by those habits and inclinations which are observed to be common to the equine species.

During the early part of this century cases of farcy and glanders were of frequent occurrence among horses, which diseases were induced by defective ventilation and drainage. The cause having been discovered, the remedy, being simple, was soon applied, and what has been the result? Why, a diminution of these maladies. But, unfortunately, even in these times, the wisdom in the construction of stables, which necessitates that expired gases shall pass out of the stable through the medium of a ventilator, is opposed by those who fill the ventilator and every aperture with straw, &c., in order, as they say, "*to keep the stable warm.*" The horse should be kept warm, doubtless, and this is easily obtained by the application of an extra rug; at the same time the air within the stable must be kept pure, both in cold and in hot weather, and this alone can be obtained by insuring effective ventilation and drainage.

In the chapter on Anatomical Considerations the present position of hay-rack, manger, and water-trough has been questioned, and reasons have been assigned why such stable furniture should be placed on a level with the horse's knees, rather than (as regards the hay-rack) above his head. In attempting to establish these facts, the organs of the horse brought into requisition during the acts of eating and drinking have been anatomically examined; and so distinctly does Nature explain why such arrangements occur, that it seems difficult to understand how such spirited firms as Cottam, Musgrave, &c., have not improved upon their original inventions by patenting extra stable-fittings more in accordance with the dictates of common sense, or, more properly, those which a superficial knowledge of anatomy suggests.

In remarks made on shoeing, attention has been directed to the position assumed by the foot of the horse during progression. For years past it has been taught that the horse brought his toe to the ground before his heel, whereas ocular demonstration and reasoning by analogy prove that during progression the horse brings his heel in contact with the ground before his toe. "The art of writing well is to know your subject," and the art of shoeing well can alone be successfully carried out by those who understand the anatomy and physiology of the foot; and for this reason the article on shoeing has been carefully revised and considerably extended, particularly in the direction of physiological details above mentioned, and of diseases likely to accrue from shoeing where Nature's wise laws have been neglected or interfered with.

Many farmers, even at the present time, when a mare is too old for work, attempt to breed from her, and what is the result?—usually disappointment, attended with loss. Like begets like, and, in the case under notice, an old mare is sure to transmit a weak constitution to its offspring; moreover, in producing horses, breeders should use great care, and bestow much attention to the selection of parents. A young and well-formed mare, if mated with a horse possessing grand points, usually produces a slashing colt. Again, sometimes defects in the female may, in the progeny, be obliterated by the good qualities of the male; but this is by no means a general rule, and, consequently, it behoves all lovers of good horses to breed from parents of sound constitution, and from those possessing perfect physical development.

The treatment of mares before and after foaling is fully considered in the article on Breeding.

Horse-breakers of the present day have not yet learnt that the horse is a timid animal. What is improperly termed pluck, is fear, which has to be allayed by kind usage, both in word and deed, when, by so doing, confidence takes the place of timidity, and the horse becomes tractable. The word " breaker" is an improperly applied term to a man who gives with firmness, yet kindness, the first lessons to a young horse, but a most fitting denomination for the person who, with spur, whip, and cruelty, breaks the temper, never inspires confidence, but drives all the best attributes out of the three-year-old with lash and heels,—qualities which as a two-year-old he possessed, and which judicious and kind schooling would have intensified.

At the conclusion of this work several pages on early training have been written with a view to prove the cruelty inflicted on racing stock by two-year-old racing. Many may ask, Where is the cruelty? Let us consider the ordeal through which a colt has to pass previously to his becoming a two-year-old performer, and then ask why, and give your verdict.

At one year and a half old this young creature is in active training; and in what state are the tissues of his body at this tender age? Why, they are growing, and consequently the muscles are tender; the bones are not fully developed, proved by the fact that the ends (*epiphyses*) and centres (*diaphyses*) of the long cylindrical bones are not securely cemented together, so as to form a strong shaft. Therefore, viewing the animal in question merely as a machine, you would pronounce it to be unfinished, and if acquainted with physiological science, would state such colt to be unformed and incapable of performing severe physical exertion; but although a baby, as it were, he has to gallop, and hard, too, with a weight on a back which ought not to carry any; and why? because the owner, by carrying out this system, expects to spend less and to gain more money. Alas! "crescit amor nummi." But should gain lead us to inflict cruelty and injury upon dumb creatures? Certainly not. Racing was instituted for the express purpose of improving our breed of horses, but does early racing bring out the best horses? It may, but it destroys nine out of every ten of them; and consequently hundreds who, if allowed more time before being worked, might grow into fine animals, are denied this opportunity, for break-down, sprained sinews, and muscular laceration unstring the harp of a thousand cords, rendering the recipients of such injuries unfit for racing purposes, and consequently from the Turf they depart, often to the cart of the costermonger.

It is contended, moreover, that short races are not calculated to produce horses of much stamina; they may, but as a rule, small weedy animals are as well able to run successfully over a short course as strongly-built ones. The latter are those who produce our best hunters, cavalry, and carriage horses, and, therefore, the breed we wish to propagate. Whereas the progeny of the former, following the rule of "like getting like," would be unable to carry a pair of boots to hounds, and consequently useless as progenitors of stock.

The great good of racing is that it acts as a stimulus, causing persons, for love of honour or money, to breed good horses; these animals, on account of their performances, are chosen as parents, and thus an improving breed is kept up.

It is needless to mention further the disasters of early racing—every horseman knows of the oft-repeated publication that such a horse is struck out of his engagements. But if no horse were trained until after three years old, then the disasters would be less frequent, and we should obtain many useful animals in the place of those ruined by early training.

"What we require," writes Admiral Rous, "is a National Prize of £5,000, to be run for by four-year-olds and upwards, three miles, which might induce horse-owners to show more mercy to young horses:" in fact, make a demand and the supply will not be wanting; take away the demand and the supply must cease. Institute those longer races for older animals and for larger stakes, and by this means horses will be preserved and their powers conserved for great feats, which powers early training deprives them of.

It is, therefore, trusted that at no very distant period such radical reform may be gained,—a step calculated, if taken, to ameliorate the sufferings of that noble creature, the horse, who contributes so largely to our health, wealth, and happiness.

JAMES IRVINE LUPTON, M.R.C.V.S.

VINEYARD HOUSE, RICHMOND, SURREY.

CONTENTS.

CHAPTER I.

The Body of the Horse anatomically considered . *page* 1

CHAPTER II.

Physic.—The Mode of administering it, and minor Operations. . 38

CHAPTER III.

Shoeing.—Its Origin, Uses, and Varieties. 75

CHAPTER IV.

The Teeth.—Their natural Growth, and the Abuses to which they are liable 125

CHAPTER V.

Food.—The fittest Time for Feeding, and the Kind of Food which the Horse naturally consumes. 160

CHAPTER VI.

The Evils which are occasioned by modern Stables. . . . 183

CHAPTER VII.

The Faults inseparable from Stables. 196

CHAPTER VIII.

The so-called "incapacitating Vices," which are the results of Injury, or of Disease 224

CHAPTER IX.

Stables as they should be 235

CHAPTER X.

A few Hints to Grooms . . . . . *page* 248

CHAPTER XI.

Points.—Their relative Importance, and where to look for their Development 268

CHAPTER XII.

Breeding.—Its Inconsistencies and Disappointments 321

CHAPTER XIII.

Breaking and Training.—Their Errors and their Results. . . . 336

INDEX 353

ILLUSTRATED STABLE MANAGEMENT.

CHAPTER I.

THE BODY OF THE HORSE ANATOMICALLY CONSIDERED.

THE horse, with proper usage, might become the companion of man. It is true, the poor animal can be goaded to excessive labour; but the creature becomes degraded when it toils beyond the sphere of mortal sympathy. No living animal should be subjected to the exactions of avarice. Life was not made to be thus debased. What, however, the horse, when properly treated, is capable of performing, remains to be hereafter demonstrated. How much it can enact, and how greatly it can benefit, when justly treated, the present customs refuse the willing drudge a chance of proving. No steed is now permitted to grow till its thews and muscles are matured. Before the season of its utility can come round, the colt is seized upon by the impatience of gain, and the baby limbs are distorted by early training, which forbids the natural powers to be developed.

We can, however, even by the inspection of the body, discover that it is admirably adapted for continuous and prolonged exertion. The maintenance of animal motion chiefly

1

depends on the provision made for aërating the blood. In
proportion as the vital current can be revivified or oxygenated,
is health promoted by those efforts, which in most bodies
would, assuredly, induce congestion and death. Age becomes
very important when the subject is thus considered. Respira-

THE FIGURE OF A HORSE, PORTRAYING THE COMPARATIVE IMPORTANCE AND THE
RELATIVE SITUATIONS OF SOME INTERNAL ORGANS.

1. The Lungs. 2. The Stomach. 3. The Colon. 4. The Diaphragm.
5. The Situation of the Bladder.

tion is in youth quicker than during adultism, because there
is so much more oxygen needed when the frame is in a growing state. By working the horse before maturity is attained,
the animal is obliged to labour before the various tissues of the
body are formed, before the respiratory organs are capable of
withstanding the effects imposed upon them during severe
physical exertion. Nevertheless, that the reader may judge
correctly of the care Nature has bestowed upon the formation of a creature destined for subserviency to man, the above
engraving is presented.

This illustration exhibits the lungs as of large proportional dimensions; while the stomach will be recognized as of more than an equally diminished capacity. Everybody must have experienced how greatly respiration is impeded by a loaded digestion; and Providence, when creating an animal destined to display speed, seems to have anticipated the probability of such a contingency. The intestines, however, are comparatively of large extent. Into these receptacles the horse's food passes, after having perfected the first process of digestion, and there it is subsequently mixed with the fluid secretion of the bowels; whereby the nutritive matter is separated and rendered fit for absorption.

The smallness of the horse's stomach is in itself sufficient evidence that the quadruped was designed to be a frequent feeder. It was not intended to endure prolonged abstinence; for almost in every region which the animal may canter over, its legitimate food abounds. Man, however, frequently starves the creature, that a loaded stomach may not interfere with the activity of the respiration; he, in his ignorance and in his presumption, not being willing to trust to the provision made, anticipatory of this accident. At other times the quadruped is suffered to overgorge, its keeper paying no regard to its requirements. After an excessive fast, a quantity of cut food is placed in the manger, and the ravenous horse eats, and eats, till its small stomach, being unequal to the reception of much bolted provender, cracks its walls from excessive repletion. Such a circumstance does not demonstrate that Nature was wrong, or that the equine races were formed unequal to their purposes; but it satisfactorily establishes, that man cannot with impunity cross the designs of, or run counter to, natural laws.

The horse was created to live off the grass of the field. This habit necessitated that much ground should be travelled before the appetite of so large a body could be appeased; and the distance was the greater as the animal was sent upon the earth

a nice feeder,—biting off the juicy tops of the herbage, not
tearing up roots and all, like the less scrupulous bovine tribe.
The time was also lengthened, by the equine race not being
gifted with a power to ruminate. The ox, having filled the
mouth, bestows little care upon the comminution of the food;
but the jaw being moved twice or thrice, thereby crushing the
herbage, so as to form it into a pellet, the mouthful is for-
warded at once to the rumen. This receptacle is large, and
is somewhat hastily filled. Then the ox retires to a quiet spot
and there enjoys its meal; the grass being regurgitated and
fully masticated, during which time the animal is said to be
" chewing the cud." The horse has no such power. The food
it gathers must be prepared by mastication and insalivation
before it enters the stomach; consequently, because of the
niceness of its appetite, and the absolute necessity for each
mouthful being separately comminuted, the horse, in a free
state, has to journey far and to feed long before it can lie
down and rest.

The equine race were meant to collect their sustenance from
the surface of the earth; and, doubtless, the tribe are most at
ease when feeding with the head lowered to the necessary
position. A dog naturally lowers the mouth when it laps a
fluid; but, if this creature be tempted to drink from a saucer,
held on a level with the ordinary elevation of the head,
repeated coughing will interrupt the draught and testify to
the inconvenience experienced by the .animal. So, in the
instance of the horse, we may infer the meal is most relished
when the head sinks to its gratification; and, to justify such
an inference, anatomy discloses a special provision made to
that end. Such a proof is of much more weight than any
assertion to the contrary of the united British public, as
emphasized by the fixed altitude of all the mangers throughout
the three kingdoms.

A serious suggestion here forces itself upon our mind; and

it is one the importance of which should recommend it to the consideration of the public. Laryngeal affections are among the most frequent annoyances of every stable, and stand foremost among the most vexatious of the many evils which the veterinary surgeon is expected to eradicate.

However, it is proved that if sustenance be swallowed, with the head at a certain elevation, it must interfere with the most

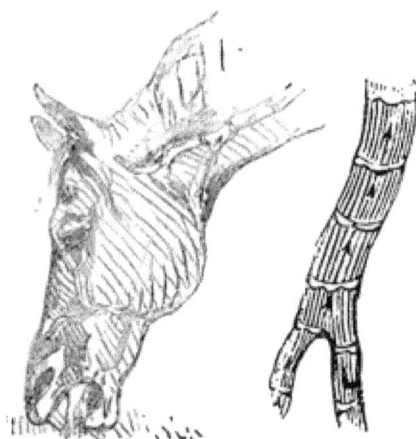

A SKETCH, DISPLAYING THE ACTION OF THE JUGULAR VALVES, WHEN THE HEAD IS LOWERED TO FEED OFF THE GROUND.

irritable organ entering into the composition of the entire body. Then, horse proprietors would do well to reflect upon the fact, and to say, how far constantly-repeated provocation may aggravate or induce the fearful laryngeal maladies to which domesticated horses are peculiarly liable.

The valves existing in the jugular veins are formed by duplicatures of its internal lining membrane; and they are so arranged as to prevent the natural tendency to regurgitate, when the fluid within the vessel moves against gravity. When the head is erect, and the venous current, flowing towards the heart, is of course downwards or is favoured by gravity, then the valves do not act; but the passage of the blood forces the duplicatures of membrane to remain close against the sides of the tube.

The jugulars conduct the dark-coloured blood from the brain; and as that important organ cannot endure the smallest pressure, some special provision was imperative to carry away the fluid, and also to anticipate the possibility of its return to oppress the sensorium. When the horse is grazing, the head is lower than the heart, and it naturally occupies that position for the greater portion of the twenty-four hours. During all that time the venous current must mount against the influence of gravitation; and to aid the reader in properly understanding the means by which this is effected, his attention is invited to the foregoing diagrams.

The elevated crest, therefore, presents a clear channel to the vital current. For that reason, the violent action or the most rapid pace of the animal never produces congestion of its brain. The racer may sink from exhaustion, but does not perish from apoplexy. The head, when depressed, however, shows the same canal divided by numerous intersecting marks. Such lines are intended to represent the venous valves, which assist the blood in its upward journey, and render impossible the slightest pressure upon the sensorium. The first thing which strikes the reader upon beholding the arrangement depicted above, is the vast number of valves; and this causes him to inquire, where was the necessity for such repeated checks. If the conservation of the brain was the only end to be attained, might not that object have been assured by a single set of valves? Such may seem a feasible objection; but to prevent the return current was, as Nature appears to have conceived, best done by repeated assistance of the onward flow; consequently, these numerous valves anticipate the possibility of regurgitation in any degree, and provide repeated checks to pressure from the supported column of heavy venous blood.

There remains, however, another provision to be explained. The return current has hitherto been spoken of, as though the upward flow of fluid was its natural tendency. Still, every person must have perceived the necessity, when liquid was to be propelled in that direction, of something resembling a forcing pump. Such an apparatus Nature has provided. The head of a healthy animal is depressed only when eating or when drinking. During the performance of either function,. muscles are contracting which compress the soft coats of the veins, and thus help to drive the circulation against gravity.

Thus, during feeding, the head is maintained in a depressed attitude for hours together; and, throughout that space, a most powerful agent is in operation. The lower jaw, while the quadruped chances to be thus engaged, is in constant motion, being opened and closed either in biting or in chewing. When the jaw sinks, the muscles of mastication are relaxed, and the venous blood rushes from the cranium into the sinuses. But when the jaw is raised by those strong motor agents, which render the bite of a horse so fearful an infliction, the current from the brain is for a moment checked, and the contents of the maxillary sinuses are energetically propelled up the jugulars. During the first half of the action, the valves are in operation, having all the strength necessary for the perfect performance of their allotted function; but, during the latter part, they are forced against the sides of the vessels by the contractive masticatory influence, and cease to act in any way upon the internal current of the blood.

Notwithstanding the strong conviction emphatically asserted by the fixed position of the nation's mangers, we must be obstinate enough to disregard human authority when we have an opportunity of studying the living book of Nature. Valves, though generally present in veins, are never discovered where the position of the vessel or other reason would render such provisions unnecessary. The Great Creator often makes one thing to serve more than one use; but never creates when *His* work can answer no profitable purpose.

The use of veins is simply that of conduits, to convey the refuse blood back to the heart, whence it is forced into the

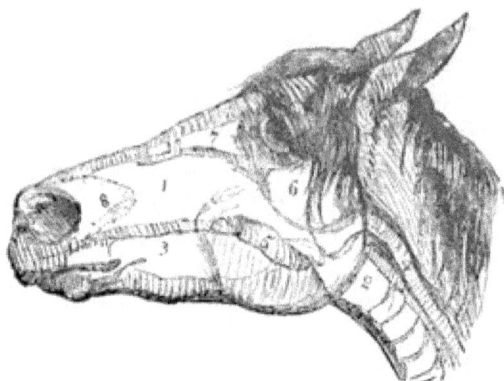

A DIAGRAM, EXPLANATORY OF THE SPECIAL PROVISIONS DISCOVERABLE IN THE
HEAD OF A HORSE.

1. The Nostril leading direct to—2, the Larynx, situated at the commencement of the windpipe. 3. The Tongue. 4. The Œsophagus, or gullet. 5. The Soft Palate, which lies upon the tongue, and affords a resting-place, whereon reposes the epiglottis, or the guardian cartilage to the entrance of the larynx (2). 6. The Guttural Pouches, or large membraneous and open sacs, containing nothing but atmospheric air. 7. Nasal or frontal Sinuses.

lungs, and there revivified or rendered equal to its many forms of nutrition. This mighty change is very simply effected. When the thorax expands, air merely enters the lungs to anticipate the vacuum, which otherwise must be occasioned by the enlargement of the chest. The air consists chiefly of two substances in a gaseous state—of oxygen and of nitrogen. The venous blood, being very near to the inhaled air within the lungs, extracts the oxygen from it, and in exchange, sends forth a quantity of carbonic acid, which is voided with the expired breath.

This change will take place when blood is extracted from the body. If the contents of some vein are exposed to the atmosphere, they will in time change from a deep Modena red to a bright scarlet hue. There is, however, this difference, which marks the two processes. The alteration, when quickened by vitality, is instantaneous; but, the change, when it ensues under human inspection, is slowly and, as it were, laboriously accomplished. The size of the equine nostrils informs us of the ample draughts of air, which the animal is fitted to appropriate; it likewise testifies to the high state of that vitality which could necessitate such a provision. Creatures with small nostrils, for instance, ox and dog, are endowed with a limited capacity as respects nasal respiration. Yet, as a recompense, such creatures are formed to inhale through the mouth. The horse, however, requires no such faculty; its nostrils are ample; and, under ordinary circumstances, the mouth is closed by a thick, fleshy screen, which hangs pendulous from the most backward portion of the bony palate.

In the previous diagram, figure 1 indicates the space allotted to the nasal chamber, near the external opening to which will be observed the numeral 8. The dotted line surrounding the last figure represents the dimensions of a blind pouch, or cul de sac, which separates the external from the internal wall of the true nostril. The existence of such a provision has long been a puzzle to physiologists; but, if these gentlemen had given *Nature* full credit for that care with which the *Common Parent* studies to preserve the beauty of the higher order of *His* children, and had considered that the horse's necessity for different quantities of air varies with different times and during different occupations, they might have sooner comprehended the utility of the development.

Where the false nostril is placed is the only portion of the nasal chamber which is not enclosed by bone; consequently, it is situated at the only place where the cavity admits of distension and of contraction. The animal, in a passive state, breathes very leisurely; at such times the nostrils would sink inward, or be deformed by the unavoidable collapse of the wall, were not the false nostril present to permit its diminution, without materially affecting the external form. But, subsequently to severe exertion, everybody must have remarked the nostril spasmodically strain, as though each effort would crack the boundaries of the opening. At such times the false nostril offers no stubborn opposition to the violence of respiration, while it serves to soften down the aspect, which, if laid bare, might show too fearfully.

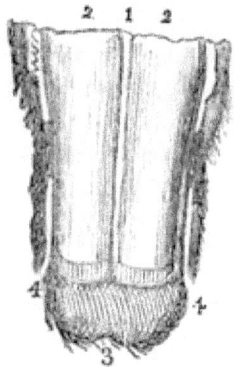

A varied capacity for admitting air also presupposes a varied capacity to alter the dimensions of the passages through which the atmosphere travels to the lungs. If the reader will again refer to the facial diagram, he will perceive a free space, in the centre of which is placed the figure 6. These spaces (one on either side of the face) represent what are termed the guttural pouches, they being merely bladders containing air, and communicating separately with each nasal chamber. A bladder with an external opening is of course most readily compressible. That no doubt may be entertained of the use for which these vacant spaces were established, they are placed immediately above the course of the atmosphere to the lungs, and would contract or dilate according to its volume.

DIAGRAM OF THE FALSE NOSTRILS.

1. The Septum Nasi. 2. The Nasal Chambers. 3. The Upper Lip. 4. The False Nostrils.

Such a condition of parts imagines the windpipe also able to alter its dimensions, so that it may be in accord with other structures ; and anatomy discloses facts which amply support such a supposition. The larynx or opening to the windpipe is composed of several pieces of cartilage and of numerous muscles. The presence of the first plastic and highly elastic structure is a proof that the larynx is of no fixed shape, while the division of the organ into distinct parts, together with the internal and external presence of many muscles of motion, is absolute confirmation that the larynx was created not only to assume various forms, but also to exhibit different capacities, according to the requirements of the animal economy.

So, also, with the windpipe itself, and the tubes which proceed from it; these are formed of distinct rings, or of separate pieces of elastic cartilage so curved as to form rings, but having free overlapping ends, which are operated upon by muscular fibre.

The diagram inserted on the next page accurately depicts such a ring; it also shows that the springy cartilage is not made of one thickness throughout, but is of that form which the mechanic of the present time recognizes as that best adapted for the preservation of continued elasticity. The shape and the free ends evince that such a ring must always have a tendency to expand, and by this perception we recognize the purpose of the muscle, which draws the extremities over each other; thus, two opposing forces regulate the capacity of the circle.

The presence of muscular fibre is always absolute proof of motion. Where muscle exists and morbid circumstances render motion an impossibility, the function being destroyed, the motor organ becomes pallid, or suffers atrophy. The existence, therefore, of such a structure in a healthy condition is always sufficient proof that the function of expansion and of contraction was present during life; thus we reach an absolute certainty that the air-passages of the horse possess a property of adapting themselves to the necessities of the animal.

Then, looking at these structures, we find them not only free, but so composed as to be always open, excepting when the momentary swallowing of the food causes the larynx to

close. To breathe is the primary necessity of life. Health
cannot be maintained unless the blood is sufficiently oxygen-

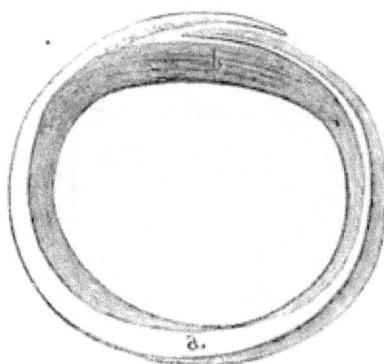

ONE OF THE CARTILAGINOUS RINGS, NUMBERS OF WHICH JOINED TOGETHER FORM THE
TRACHEA, OR THE WINDPIPE OF THE HORSE.

a. One of the cartilages from the trachea of a horse, having free and overlapping extremities.
b. The muscular fibre situated within the ring, which regulates the diameter of the circle.

ated: this fact makes us doubt the national wisdom, which
persists in thrusting the quadruped into stables, rendered
close and hot by the products of impurity.

In the head of the horse we discern evidences of the care
bestowed to preserve a harmony of form. Above the nasal
chambers are certain hollow spaces, indicated by the figure 7.
These empty chambers may serve to impart depth to the
voice, but as the horse is generally a silent creature, such
obviously must be only a secondary purpose. To preserve
the undulation of the outline was assuredly the primary intent,
though, at the same time, the vacancies aid the reverberation
of sound, and, with other structures, also lighten that part of
the body in which they are situated.

The passage of the air to the lungs, and the admirable
provisions to admit its ingress and its egress, without destroy-
ing the mild and characteristic aspect of the quadruped,
having been described, it now becomes our duty to dwell
upon the extraordinary conditions which conserve the passages

of the food. Referring again to the diagram here reproduced, we see the mouth, occupied by the tongue (fig. 3), on the base of which organ reposes a dark body, particularized by the figure 5. This last is the soft palate, which drops pendulous from the osseous roof of the masticatory orifice. Upon the soft palate lies the most forward of the laryngeal cartilages, which is anatomically spoken of as the epiglottis; while the most backward of the laryngeal cartilages, which are called the aretenoids, repose beneath the roof of the pharynx. This pharynx is the enlarged and muscular commencement of the gullet, the situation and direction of which channel is notified by the number 4.

We thus perceive in its course the food is apparently thrice

A DIAGRAM, EXPLANATORY OF THE SPECIAL PROVISIONS DISCOVERABLE IN THE HEAD OF A HORSE.

1. The Nostril leading direct to—2, the Larynx, situated at the commencement of the windpipe. 3. The Tongue. 4. The Œsophagus or gullet. 5. The Soft Palate, which lies upon the tongue, and affords a resting-place, whereon reposes the epiglottis, or the guardian cartilage to the entrance of the larynx (2). 6. The Guttural Pouches, or large membraneous and open sacs, containing nothing but atmospheric air. 7. Nasal or frontal Sinuses.

forbidden to enter the gullet of the horse. In the first place, there is the soft palate, retained firmly in its position by pressure of the epiglottis; the second obstacle we recognize in the opening of the larynx; and the third impediment appears in the arotenoids, that seem to bar all entrance to the tube which leads to the stomach. Moreover, the gullet itself being a muscular organ, in the passive state of semi-contraction is closed; thus appearing to oppose a further hindrance to the admission of sustenance into its proper receptacle. However, upon inquiry, the reader will discover these provisions, which appear at first glance to be ranged against the entrance of nutriment, are in reality only so many elaborate protections, all tending to the comfort and well-being of the animal.

The soft palate so effectually closes the posterior of the mouth, as to prevent that cavity from being employed to modulate the voice; though such a peculiarity does not distinguish all the equine tribe. Everybody must have remarked the bray distend the jaws of an ass, whereas, the neigh flutters only the nostril of the horse. The different channels through which the sound has to emerge, fully accounting for the marked contrast which is conspicuous in the voices of the animals. Moreover, the horse does occasionally vomit; but, save when the organization is disturbed by the agonies of death, the voided matter is generally ejected through the nostrils.

However, the reader will, perhaps, best understand how the apparently closed cavity is rendered subservient to its uses by the process of deglutition being described. A portion of food is bitten off by the incisors; the substance is by the action of the tongue next passed to the molars, or is placed between the grinding teeth. There it is thoroughly comminuted; while this is being performed, the saliva is secreted and mingled with the mass so as to render it quite soft or pultaceous. In this state it is formed into a pellet, and is then pressed by the tongue against the palate or roof of the mouth. The morsel, being now round and soft, is afterwards, by a more energetic contraction of the tongue, driven against the pendulous palate, which seemingly closes the posterior of the orifice.

The last organ lies in that direction which enables it to offer a formidable resistance, especially when supported by the base of the tongue, to any substance proceeding *from* the stomach. In the contrary direction it is only held down by the epiglottis; that comparatively feeble body is forced to yield before the greater contractile power of the lingual organ. The epiglottis flies forward, covering the opening to the larynx, in which position the posterior cartilages or the aretenoids also fold over the more forward protector. A secure floor is thus formed, preventing anything from falling into the windpipe, where intrusion of the smallest substance would provoke the most alarming spasm; while a roof to the passage is also made by the raised soft palate, whereby the nasal chambers are protected from the encroachment of undigested matters.

A safe way being thus provided, the pellet is shot into the pharynx, which, independently of the will, immediately contracts upon any substance coming within its reach, and drives the morsel into the œsophagus, or gullet. The tube, surprised by the presence of the morsel, is obliged to separate for its reception; but it immediately closes on the stranger, thereby driving it lower down, when the contractility of the fibre being again aroused, it is once more driven onward; and this action is continued until the food is safely lodged within the walls of the stomach.

Few persons can comprehend the above explanation without being forcibly impressed by the beauty and the nicety of the whole arrangement. The elevation of the soft palate closes the nostrils, and at the same time provides a floor for the gaping passage to the lungs. The motion of the soft palate nudges the epiglottis, which lies upon it and causes that cartilage to bend over the opening to the larynx. The bowing down of the epiglottis induces the aretenoids also to stoop, thus forming a safe floor to the necessitated passage. Across the chasm, now rendered secure, the food is shot into the pharynx and conveyed to the stomach, the whole process being accomplished in an instant, for the act of swallowing provokes no sensible impediment to the continuance of respiration.

These things, however instructive or amusing they may be when related, nevertheless are too little thought of; nor is the horse itself sufficiently considered. Were the lessons, which its body should teach mankind, properly understood, those abuses, that are at present limited to no class, would instantaneously cease to be practised. Most people of this country, however, treat the horse as though it were an original inhabitant of the English climate. Rich and poor in this respect are equally faulty; save that those are most to blame who, possessing wealth, can command the leisure requisite for inquiry, and, being blessed with ability to gratify their inclinations, have no excuse for lack of sympathy in the pressure of necessity.

The horse carries about its person the signs which testify he once roamed within a warmer climate than our northern region. The certificate of his origin is legibly written in the

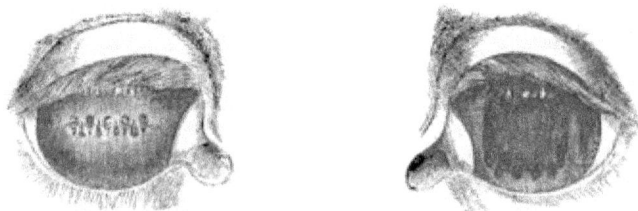

THE PUPIL OF THE HORSE'S EYE IN THE OPPOSITE STATES OF CONTRACTION AND DILATATION, SHOWING THE SITUATION AND THE USE OF THE CORPORA NIGRA.

eye. This organ mutely attests that the temperate zone was not the birthplace of his progenitors. He has long been a captive in Britain, but the proof of his proper dwelling-place no time can obliterate. The eye of the horse, like that of the camel, displays a special provision, fitting the creature to endure the strongest glare of a tropical sun, even when reflected from a level waste of shining sand.

The corpora nigra, in the eye of the camel, are black bodies, pendent from the margin of the iris. The purpose of so special a provision is not apparent, when darkness occasions the opening to dilate; but when the glare is powerful—so powerful as to induce blindness even in the natives of those lands where a concentrated light is possible—then, the wisdom of this beautiful disposition of parts becomes apparent.

The pupil of the horse's eye is never circular, being, when much dilated, rather oblong in figure; but, when exposed to the direct rays of the summer's sun, the opening energetically contracts. Then the pupil is best represented by a mere line, for the edges of the iris at such a season seemingly touch each other. In this condition the uses of the corpora nigra can hardly be mistaken: the little black bodies appear to fit into one another, forming apparently an impenetrable network opposed to the entrance of too strong a glare.

Let the reader, however, temperately consider this matter. The pupil in the eye of the horse is not more distant than two inches from the origin of the optic nerve. When the division to be seen through is so close, and the object to be viewed is exhibited under the strongest natural light, the merest crevice will be equal to all the purposes of perfect vision. The full glare of the sun alone occasioning the horse's pupil to contract, that which causes the opening to almost shut also provides the excess of light, which alone could render useful that narrow division through which objects must be recognized, while the dark bodies being stationed before the point of sight, answer the purpose of the smoke which lads load upon glass when they are ambitious of gazing at the sun.

The reader must have remarked the pupillary line, through which the domestic cat exercises perfect vision, during the bright noon of a Midsummer day. The eye of the feline race

is, however, possessed of no other protection. The contraction
may be the effect of weakness of sight ; at all events, the
revisor thinks he may conclude the far-famed eye of the cat
to be inferior to that of a horse. The domestic mouser is
popularly said to see in the dark ; the steed has been long
known to penetrate the gloom which sets the strained vision
of its master at defiance ; but it remains to be granted that
both horse and cat are equally fitted to roam by night. The
habits of the herbivorous creature would, however, assert it
to be possessed of such a faculty ; and the anatomist discovers
in the visual organ of the animal a provision specially adapting
it for these nocturnal peregrinations.

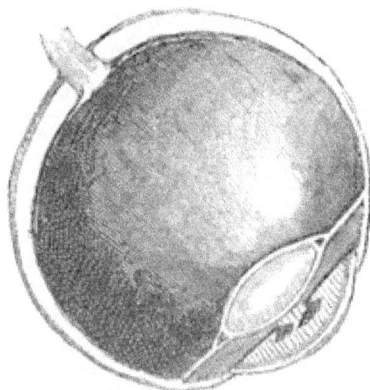

DIAGRAM, EXPOSING THE INTERIOR OF THE HORSE'S EYE, AND DISPLAYING THE SITUATION
OF THE TAPIDUM LUCIDUM, OR GLOSSY SURFACE DEVELOPED WITHIN THE ORGAN.

Upon the upper and forward surface of the inner, dark
chamber, and so placed as to catch, to concentrate, and to
reflect every stray ray of light upon the optic nerve, the
tapidum lucidum is discovered within the globe of the horse's
eye. This structure is, after death, very bright, or of metallic
lustre, and, because of its concave form, is admirably adapted
to its particular function. That no doubt may remain as to
the design of such a provision, the tapidum lucidum is found

only within the eyes of those quadrupeds created to roam by night. It is altogether absent in such animals as were destined to move about during daylight.

The tapidum lucidum, therefore, viewed in conjunction with the corpora nigra, becomes an inferential proof that the horse originally inhabited some land in which the coolness of the night offered the greatest temptation for pleasant pasturage; consequently, this creature was formed to enjoy the bounties among which it was permitted to roam. We know the cat was imported from the tropics; and, seeing that the eyes of both animals, in one marked particular, resemble each other, we may conjecture the horse originally inhabited a warmer climate, while the likeness between the equine race and "the ship of the desert" demonstrates that locality was the hottest portion of the earth.

The eye of the horse is also provided with a power which could seldom be needed in these northern climes, where the fleetness of the equine tribe might readily set at defiance the comparative feebleness of all the predatory beasts of prey. Besides, the wooded state of this country must have rendered the presence of telescopic vision unnecessary. Upon the far-stretching level of the desert, however, where larger and more ferocious animals prowl by night, the possession of such a faculty would be a needed protection. Accordingly, we find the interior of the globe to consist chiefly of water, the outward covering being formed of a tough substance, which is easily compressible, while all the hidden portion of the exterior is enveloped by muscular fibre.

Situated directly upon the forward portion of the ball, are the two oblique muscles. These are inserted at opposite places, and each pulls in a contrary direction to the other. The two, simultaneously acting, could not move the organ, but would, obviously, tend to fix it, or to render the globe stationary. The outer substance of the horse's eyes is composed of a thick and pliable covering, purely tendonous in character. The interior consists of fluid, perfectly pure and transparent. At the back of all is placed the optic nerve, while the exterior is enveloped by several thick and straight muscles.

The motor agents are endowed with an ability to contract or to shorten in their reach. When parts of this nature operate upon a plastic substance, which is filled only with a fluid, they must of necessity tend to alter the shape of that body on which they repose. The oblique muscles act to prevent rotation; the pressure, therefore, can only compress,

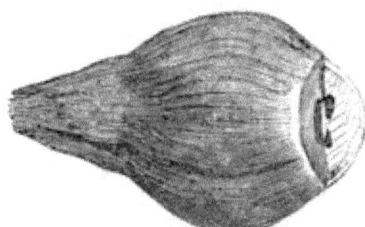

DIAGRAM, DISPLAYING THE COATING OF MUSCULAR FIBRE WHICH COVERS THE SOFT GLOBE OF THE HORSE'S EYE.

elongate, and force backward the ball of the eye. By such a capacity that telescopic property is produced, which man feebly imitates by a complex and costly machine.

Anatomy also discovers another important function proper to the eye of the horse, which equally indicates a sandy plain to have been the original habitat of the tribe.

Horses, when travelling in the desert, are compelled to perform their journey in a cloud of dust sufficiently dense to blind them, had not Nature supplied to the horse's eye a protection against so terrible an affliction, which consists in the existence of a membrane, which is reflected over the front part of the eyeball, and thence over the interior of the eyelids. If the layer common to the eyeball or eyelid be touched by any foreign matter, it is incapable of communicating to the sensorium more than a feeling of uneasiness; but, when the two layers, viz. that of the eyeball and eyelid, meet, the offending agent, i.e. when the two surfaces of the membrane are brought into opposition, then agony ensues.

The membrane, now under consideration, renders it an impossibility for any substance to get "into the eye"; the pain present, when such an assertion is commonly made, gives the strongest proof that the foreign body is retained between two surfaces of that delicate structure, which is called conjunctiva. Dryness is, however, destructive of the feeling and of the transparency of this membrane. Nature, therefore, has created a special gland for assuring its perpetual moisture. This last body is situated immediately beneath the surface, under the upper lid, and towards the outer corner of the eye. It is, on ordinary occasions, stimulated to send forth its secretion by the act of winking, and the outer corner, being situated above the inner

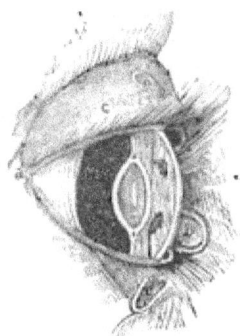

DIAGRAM, EXPLANATORY OF THE SITUATION OF THE LACHRYMAL GLAND, OR THE SOURCE OF TEARS, AND OF OTHER PARTS PROPER TO THE HORSE'S EYE.

a. The outer angle. b. The puncta lachrymalis, or round body, situated at the inner angle of the eye. c. The lachrymal gland, covered by the upper lid, and placed near the outer angle of the eye. e e. The position, extent, and doubling of the conjunctiva, or of the investing mucous membrane, which envelops the outer portion of the globe, and lines the lids.

corner of the horse's eye, the moisture is, by the motion of the lid, instantaneously brushed over the circular globe.

The gland of the horse, however, has a distinct use, not shared by any similar provision to be found in the eye of man. In the human being, grief or pain provokes the secretion : these are always accompanied by floods of tears. Some writers assert, they have witnessed agony induce tears in the quadruped; but the reviser has seen fearful operation inflicted on the noble animal,—he has heard groans testify to the sufferings endured;—yet, he has never beheld tears flow from the eyes to such an extent as to even approximate to weeping.

Pain, when occasioned by some foreign body between the two layers of membrane, produces not weeping, but a positive overflow of liquid, the purpose of which will be best explained after the reader has been made acquainted with a particular organ, situated at the inner angle of the eye.

The lower corner of the organ is characterized by a round body, which, being enveloped in a single layer of membrane, is strictly without sensation. Upon this body the grime of the human eye accumulates, and we shall shortly perceive that

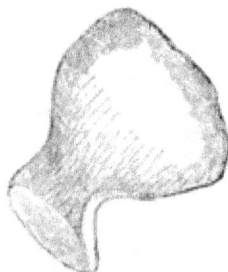

THE CARTILAGO NICTITANS, WHEN RE-
MOVED FROM THE EYE OF A HORSE.

its presence in the horse is not without a purpose. Next to the foregoing development, and so placed as to accurately fit the globe, is a structure which anatomists name the cartilago nicitans, or the winking cartilage. The forward portion of this cartilage possesses a fine edge, while its base presents a broad surface, which reposes upon the fat at the back of the orbit. Now, as fat is not compressible by ordinary force, whenever the muscles draw the globe backward, the adipose matter is driven forward: this last carries with it the cartilago nictitans, which is consequently projected suddenly over the surface of the globe. But, when the muscles relax, the fat resumes its original place, and with it the cartilage also retires.

When any foreign body gets between the two layers of membrane, instant winking results; the gland, stimulated by the motion of the lid, sends forth a gush of liquid. It is not simply a tear or two, but a deluge of fluid is emitted; this

flood, aided by the action of the lid, carries the foreign substance in the course of gravity, or from the external towards
the internal corner of the globe. While this is taking place,
the pain also excites the powerful muscles of the eye to
spasmodic activity. With every spasm the fat is displaced,
and the cartilage darts from the inner corner partially over
the round surface of the eyeball. The process continues until
the substance is partly brushed and partly washed to within
the range of the fine anterior edge of the cartilage, when, by
its withdrawal, the foreign particle is lodged upon the round
insensitive body developed at the inner corner of the eye.
Towards the last point the tears naturally tend, and any
exciting substance, when there placed, is soon floated on to
the hair of the cheek.

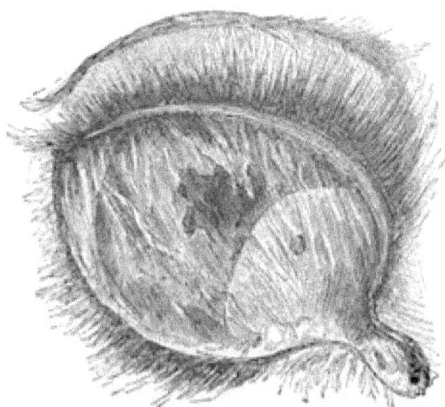

DIAGRAM, TO ILLUSTRATE THE ACTION OF THE CARTILAGO NICTITANS UPON
THE HORSE'S EYE.

By joining these many proofs, we gain a moral certainty
concerning the region whence the horse originated. The eye
is seen to be gifted, not only with a special provision against
the glare of the desert, but it also possesses a peculiar development fitting the animal to enjoy the cool pasturage of the
night. The eye is likewise endowed with a telescopic power,

suited to sweep the far-stretching horizon of the sandy waste. Moreover, the organ discloses a special apparatus evidently designed to overcome those accidents to which inhabitants of arid plains, when rapidly travelling long distances, and in large herds, were exposed.

The reviser is aware that gentlemen of known probity have reported the existence of herds of wild horses careering free and unbroken over the plains of Asia. Such was formerly said to be the case, and was also credited as an established fact, with regard to Southern America. Subsequent inquiry, however, has shown that the wild animals of the pampas are no more than neglected flocks roaming, apparently without an owner, but which, in reality, are allowed thus to gain a cheap livelihood by a careless proprietor. These American herds are liable to the claim of some man, almost as wild as the animals themselves; so, also, the reported Asian quadrupeds turn out to be the recognized possession of some wandering Tartar.

SOME OF THE DEEP-SEATED MUSCLES IMMEDIATELY INVESTING THE SPINE OF THE HORSE.

1. The hair. 2. The skin. 3. The adipose, or fatty tissue directly under the skin. 4. The bursæ mucosæ, or synovial sacks placed above each dorsal spine. 5. The yellow, elastic ligament connecting the dorsal spines together. 6. The spines of the dorsal vertebræ. 7. The semi-spinalis dorsi muscle. 8. The heads of the ribs. 9. The levatores costarum muscle. 10. The ribs.

However, to leave the consideration of particular parts, and to view the entire body anatomically, the vertebræ or spinal chain, as forming the base of the skeleton, becomes of primary

importance. The backbone of the horse consists of various
pieces, so firmly held together by interlacing ligaments and
muscles, that students, when desirous of dividing the spine of a
dead animal, often find it easier to saw the bones asunder than
to separate them with the knife. The neck is composed of
seven bones; the back is formed by eighteen vertebræ; the
loins consist of six pieces; and the sacrum is made up of five
distinct parts, although long before adultism all of these last
are united by osseous junction.

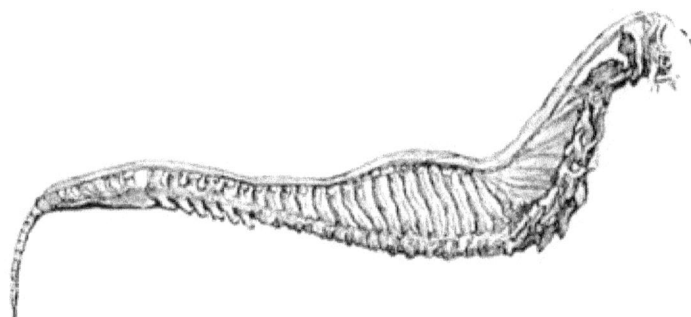

The sacrum, therefore, is not reckoned among the true
vertebræ, the number of which, however, amounts to thirty-
two. Of these many divisions, the bones of the neck alone
are not subject to deviations. The lumbar may be five or
seven, and the dorsal limitation is either one above or one
below the usual amount, neither of which varieties are of very
rare occurrence. The links of the backbone differ in form
and in function. The dorsal vertebræ seem, at first sight, to
possess no lateral processes, whereas, in the lumbar region,
these developments are so extended as to constitute the principal features of the several parts. So, also, the two first bones
of the neck enjoy great motion, and all the links of the neck
are very far from stationary. But the parts of the back, on
the contrary, are all but fixed; yet, although each is endowed

with a very limited movement, the whole is gifted with an evident elasticity, which affords an easy seat to the rider.

Along the top of the backbone runs a strong cord of yellow elastic fibre, which unites the several parts, holding these firmly together as one whole. The elastic cord, however, passes directly from the last dorsal spine, to be fixed into the back portion of the skull, thus skipping over all the bones of the neck. The fibres of this cord are longitudinally arranged, and, however elastic such a substance may be, the dorsal arrangement would not allow of that freedom of motion which was requisite in the neck of an animal which was to crop its food from the surface of the earth.

The necessity, however, was fully met by an elastic cloth being, as it were, thrown over the cord, and extending thence to the bones of the neck. By this arrangement frequent attachments were avoided and grace of outline was preserved, while no deterioration was made in that provision, by means of which the heavy head is supported without apparent strain upon the muscular fibre. One end of the elastic expansion being inserted into the cervical bones, all the ease and beauty of movement is rendered possible by the retractile property of the cloth-like ligament being fully equal to the sustenance of the weight, but not strong enough to resist the action of the muscles when excited. Thus, the muscles situated at the base of the neck serve to depress the head; the elastic cloth answers as a counterpoising force, which steadies the movement; the action of the motor agents near the crest, aided by the ligamentous elasticity of the neck, serve to elevate the part, while the muscular power at the base of the bones regulates and guides the upward motion.

But the reader may be desirous to learn how far the back of the animal is suited to endure the weight of the rider. The bones of the spine, not being joined by osseous union, may give solidity to the part; but it must be self-evident the chain possesses no inherent power to sustain the smallest pressure. Therefore the body of the rider, when placed upon the back, cannot be upheld by bone alone. The weight must repose upon the muscles and the ligaments by which the solid parts are kept together. Man, therefore, when mounted upon a horse, is seated upon elastic substances, animated by the powers of vitality; this circumstance readily accounts for the pleasurable feelings and the lightness of spirit communicated to the master when in the saddle; although the delicacy of the structures on which the burthen is cast should also instruct, that an elaborately and a delicately organized body ought to be shielded from labour until age has confirmed and strengthened the several portions of the frame.

When contemplating the uses for which the quadruped was created, we perceive the necessity of that huge mass of muscular fibre with which the back is cushioned. We also recognize the beauty of intention which those numerous sup-

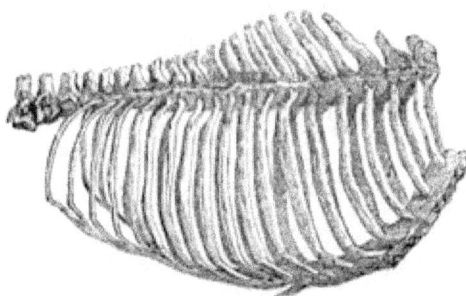

THE THORACIC FRAMEWORK OF THE HORSE.
Showing the manner in which the ribs spring from the spine to unite upon the bone of the breast.

ports called ribs, embody and declare. These props, eighteen on either side, must greatly strengthen the main structure, although each is of a loose texture, and every one is more or less pliable. The innate property of elasticity belonging to the horse's ribs seems to have been long known to country urchins, who, out of these bones have been accustomed to form bows, whence to propel juvenile arrows. Nature, however, seems not to have been satisfied with this provision, for the inferior portion of the ribs consists of cartilage, which anatomists speak of as the most elastic substance in the body; this yielding termination rests on the sternum, or breast-bone, as structure, more than three parts of which are composed of the last-named material.

The manner in which the fore limb is united to the trunk, likewise offers matter for the reader's admiration. Considering that the horse is a beast of burthen, man, were he designing a creature fitted for such uses, would assuredly have sought to gain strength by the insertion of bone. Bone, however, would

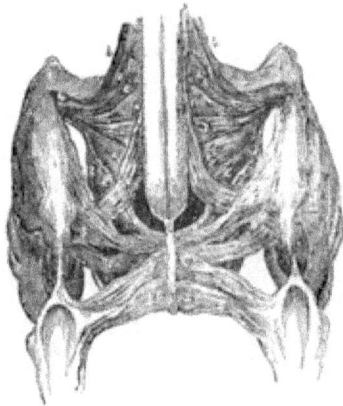

SOME OF THE MUSCLES WHICH ATTACH THE FORE LIMB TO THE TRUNK.

Three muscles have already been removed, viz., the panniculus carnosus, the levator humeri, and the latissimus dorsi.

1. The Trapezius. 2. The Serratus magnus. 3. The Subscapulo Hyoideus. 4. The Rhomboideus. 5. The Pectoralis anticus. 6. The anterior portion of the Pectoralis magnus. 7. The Pectoralis parvus. 8. The Pectoralis transversus.

have interfered with that agility which, no less than strength, is an attribute of the horse's body. The presence even of a clavicle joining the shoulder to the thorax would have exposed a jumping quadruped to repeated fractures. Nature, therefore, bound the parts together by interlacing fibres. And to afford an idea of the marvellous care bestowed on this arrangement, the foregoing diagram is submitted to the contemplation of the reader.

The rider, therefore, when mounted on a horse, is not only seated upon fleshy and ligamentous fibre, and upheld by pliable bone based upon elastic cartilage, but, as the thorax is supported by the anterior extremity, he actually swings upon the strongest and most yielding substance known throughout animated nature. Could mortal ingenuity, by the exercise of any force or duration of thought, have perfected so exquisite a work?

The bones within the fore limb are not self-sustaining. Remove their coverings, and they will not retain their several places, but will fall in a heap upon the earth. The fact proves that the osseous framework, although it confers solidity upon the body, is, nevertheless, upheld by the structures with which it is enveloped. The bony column, however, when united and bound together, exhibits an intention of bestowing elasticity, quite as much as of conferring strength. In the first place, the solid column is crowned by a broad, but thin plate of cartilage, the yielding property of which has already been dilated upon. So that the trunk not only swings upon living fibre, but the primary weight is endured by what anatomists designate "the most elastic substance in the body," of a shape and form which develop to the uttermost its bending property.

The arrangement of the shoulder-blade and the bone on which it rests being angular, evidently contemplates a yielding to any force coming from above. The two next bones cannot be viewed as meant solely for strength, though the several parts of the knee and shank are slightly columnar in their order; nevertheless, the pastern bones again display an intent to yield rather than a design at gaining decided resistance. Yet, even there remains further food for contemplation when viewing these dry bones of a quadruped. The shock, of which

the rider complains, when doomed to cross the trunk of some
poor animal whose body has been disorganized by abuse, is
occasioned by the bones having been, through disease, thrown
from their natural positions.

Engineers well know that sand will oppose the force of a
cannon-ball, the power being rapidly exhausted, which has to
travel through numerous separated particles. Each grain
of sand, therefore, being distinct, a bag of that substance
offers a good preventive to the concussion produced by the
explosive force of gunpowder. But the
reader, when endeavouring to ascertain
the provisions instituted by Nature to
save the equestrian from concussion,
can at once perceive the purpose for
which the osseous support of the limb
was formed of several pieces, as well as
appreciate the beauty and grace of motion which are thereby assured.

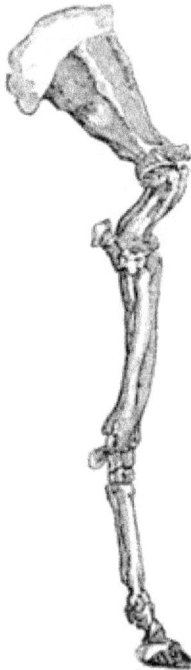

Looking at the illustration, we observe
that certain of the component solids of
the limbs are altogether out of the perpendicular, and, consequently, must
receive other support than is derived
from the bone immediately below them.
Indeed, no portion of the structure is
decidedly columnar in its arrangement.
Either the parts are crooked, or they
lean in a direction from the plummet-
line. The angularity of the two topmost
THE BONES OF THE FORE LIMB. pieces can, however, not possibly escape
notice; neither can the slanting position
of the pastern bones fail to attract attention. Noting these
peculiarities, the reader recognizes that the bones of the fore
extremity cannot be self-sustaining, but they must be upheld

or retained in their relative situations by the structures which surround them in the living subject.

The scapula and humerus, or the two topmost bones, are rendered firm by the joint action of the powerful extensor and flexor muscles, appertaining to the shoulder. The pastern bones transfer their weight to the strong tendon which passes immediately under their lower surfaces. The other bones are held in their situations by the energetic contractility of the muscles which embrace them. Hence, it is obvious, the rider, when seated on the back of a horse, is not upheld by any osseous resistance. His burthen reposes upon living fibre. The bone limits the sphere of contractility, and thus gives firmness to the limbs; but it endures no portion of the weight. So exquisitely has Nature adapted her creature to its uses, that in the horse man is provided with a means of conveyance, remarkable for fleetness, but more wonderful for the elastic and buoyant seat which an admirable body affords.

Had direct weight been cast upon the bones, the shock communicated by placing the foot upon the ground would have been so concussive as to have made the saddle a seat of torture. This is no speculative conjecture, but it is a deduction drawn from positive fact. Hard work causes the pastern bones to quit their natural slant, and to assume a more upright direction. They very rarely become actually perpendicular; but as they verge to a vertical direction, the jar communicated to the

THE PRINCIPAL FLEXOR TENDONS OF THE FORE LEG.

1. The Perforans. 2. The Perforatus. 3, 4. Accessory Muscles. 5, 6. Restraining Ligaments. 7. The Pedal Cartilage divided. 8. The Navicular Bone.

rider becomes most distressing. The tendons of the foreleg are, therefore, of all importance; the utility of these structures cannot be better illustrated than by appealing to the terrible effects which ensue upon injury to these organs.

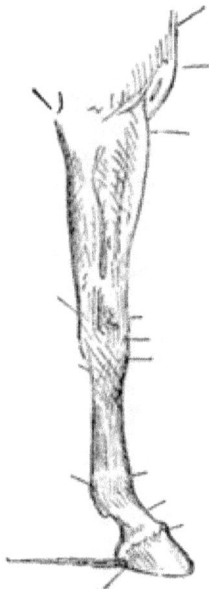

However, that the reader may fully appreciate the simplicity developed in the various arrangements exposed upon dissection, the next illustration is inserted, against which numerous lines are fixed. Those marks indicate the points where synovia, a substance like to white of egg, is interposed between the extremities of the bones. Each separate bone, thus, not only rests upon a liquid, but the ends of these formations are, likewise, tipped with cartilage, thus doubly securing the ease of progression; for concussion of the foot would be likely to occur, had not the above wise arrangements necessitated that the shock, given to the foot when it first reaches the ground, should travel through different bones, tipped by cartilage, and separated by the interposition of a fluid.

THE ARTIST'S IDEA OF A HORSE'S FORE LIMB.

The lines indicate the places where synovia (or a fluid resembling white of egg) is interposed between the different structures.

Looking toward the quarters of a horse, we perceive that the spines of the loins (*lumbar vertebræ*) bend forward, while those of the haunch (*sacral vertebræ*) incline backward, thus leaving a free space dividing the uppermost bones of two neighbouring regions. Inspecting the last lumbar bone, we ascertain it to be united by its lateral processes, yet it does not touch the first sacral's body, all other parts of the chain joining at their centres.

What takes place at this spot which could render imperative

such an arrangement? In what action is the inclination of the trunk so opposite to the position of the quarters as to render necessary such a special provision as is here exemplified? In prancing, in rearing, and in jumping, the hind legs are firmly planted upon the earth; then, by exertion of the powerful muscles of the quarters, the forward trunk is raised. This action could not have been exhibited had the spines of the sacral bones ranged in the same direction as those of the lumbar vertebræ; and to enforce the reason of this evident provision, a free space characterizes this particular joint, others being formed by the interposition of cartilage.

The skeleton of the quarters is characterized by further distinctive peculiarities. The sacral bones are fixed one to another, and joining them at the spine, is the huge hip-bone. This is the heaviest of the many weighty pieces which compose the osseous frame of the horse. It is irregular in form, and remarkable for an unusually rugged exterior. An anatomist, by simply inspecting it, could designate its uses; so emphatically is everywhere written the origin and insertion of

THE BONES OF THE HIND EXTREMITY.

powerful motor muscles. In every ridge, in every indentation, in every inequality, anatomy discovers such a purpose; thus, when "the gnarled and bossy" developments upon this bone are viewed in conjunction with the solid and uneven appearance of the lower osseous supports of the hind limb, no person, properly instructed, can doubt that the quarters are peculiarly

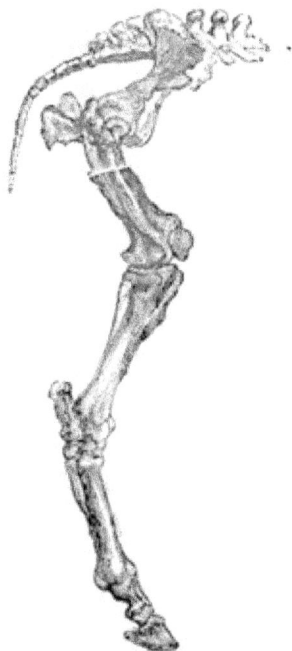

3

the seat of muscular power in the equine race, and that the angular arrangement of the bones suggests the immediate purpose of flexion and extension.

THE HIND QUARTER OF A HORSE, FROM WHICH THE SKIN HAS BEEN REMOVED.

The inferior bones of the annexed sketch lead to the foot; but as the osseous structure of this part is illustrated in a previous sketch, and as the fore and hind feet of the horse are in the leading particulars alike, little need be written concerning them, with the exception that the hind foot of the horse being the point whence all the strain of propulsion must proceed, the part, from such a cause alone, will be liable to certain distortions. The evils engendered by the cruel impatience of mankind, which forces the colt into too early labour, causes the natural position of this member to become distorted. The pastern bones, should the toil continually be enforced, assume a vertical position, and the shank bone afterwards projects. If these warnings are disregarded, inhumanity provokes the heels to be drawn upward, and a valuable helpmate is thus incapacitated from assisting man in his earthly task.

While writing of the horse, it should not be forgotten that in this country there is another animal—the ass—which properly belongs to the equine race, and which is liable to most of the evils, as well as worthy of much of the commendation already pronounced for the horse.

The donkey labours beyond the care of its enslaver, and without the region of human sympathy. It is, however, chiefly the property of those whose feelings are subject to

their necessities, when its food is the refuse of the stable, but
its work is often as despotically severe. Be its toil exhaustive,
let it work without cessation throughout the day and far into
the night, no eye regards its fatigue with commiseration. It
is an object only to laugh at. The popular belief is, that the
tribe is so peculiarly hardy, as to be altogether removed from
the necessities, the liabilities, or the accidents common to every
other form of life.

The young horse may be stinted in its food, but it is spared
from work until a certain period has arrived. All classes have
their stated ages when the colt should first begin to labour ;
but the ass has no recognized season of rest, even for its immaturity. It is forced to work so soon as need can see in the
growing body a capacity to assist. Foals are often to be seen
dragging loaded trays about the streets of London, and the
day's toil is lengthened or shortened by the quickness or the
slowness of the day's sale. The food is, during this time, the
refuse of the stock; seldom can the owner spare from his
earnings that which will purchase fodder for the life which is
the partner of his fatigues.

The donkey is harnessed for the early market. Tho costermonger rides with his family to make his bargains for the
day, and the stock-in-trade being procured, he and they ride
with it back again. The wretched creature, on reaching home,
is not released or permitted to rest, but has to support the
tray whilst the family arrange for the display of vegetables
which are intended for the morning's sale, after which the
poor brute is started on his daily round with a kick or blow,
accompanied with continuous threats, which are usually carried
into effect.

If the proprietors of asses have few faiths, they are all
thoroughly imbued with one belief, which is, that the animal
in their keeping cannot possibly feel exhaustion. Their
credulity does not stay here. They are impressed with a con-

3 *

viction that no creature of the donkey tribe has any sort of feeling. The last article of belief makes the man select the weakest portion of his dumb servant's spine for a seat, when he feels inclined to ride. The reader, to whose notice diagrams of the equine spine have been submitted, knows that the loins alone are unsupported by other bones. The absence of that which renders this region the weakest division of the vertebræ also makes this portion of the quadruped's back the most yielding and elastic. Here, the fashion of vulgarity fixes the rider's seat, when he strides the ass. The veterinary student will remember that few of the lumbar bones in the carcasses he dissected, when at college, were in their integrity. The reviser has encountered two, three, and even four bones of the six, which compose the part, locked together by osseous deposit.

Such a form of union proves the animal to have suffered inflammation. The injury must have been endured, and the agony must have run its course, for an osseous junction is positive evidence that all the stages of inflammation have been survived. Few persons, when they behold the young donkey stagger under the weight of its six-foot rider, care to think of this; nay, the writer has beheld really worthy gentlemen stand and enjoy the scene of activity presented at evening time by a rural gipsy's encampment. The women were laughing, the men were shouting, while the more jovial of the gang were racing on the common. Those poor donkeys, which already had been goaded to the performance of no ordinary day's toil, were carrying terrific loads, and beaten till they galloped, despite the deep-seated anguish with which they were afflicted.

The reviser can further testify, that among the scores of carcasses which he has dissected, he never examined the body of a donkey, however young it might have been, that he did not encounter appalling proofs of internal injuries,—injuries which had resulted in change of structure, and which would have consigned the horse to the knacker's yard. Yet, the animals thus maimed were working up to the date of purchase; the inability to move was attributed to the obstinacy which is generously supposed to characterize the ill-treated animal, and the blows fell heavier in proportion as its actual condition should have appealed to human forbearance.

The country is not secure, the people are not released, from barbarism. Who can contemplate without remorse, and, alas! " miserabile dictu," sometimes with a smile, the sufferings of that much-abused animal, but willing slave,—the donkey ?

CHAPTER II.

PHYSIC : THE MODE OF ADMINISTERING IT, AND MINOR OPERATIONS.

IT is among the worst features of modern society that, while it boasts of several worthy gentlemen who can draw largely upon their bankers, there are in its ranks so very few who would willingly submit to the smallest personal exertion for the fulfilment of that which they confess to be a moral duty. Would these individuals occasionally lounge towards the stable, the cost of its maintenance might be decreased, and, nevertheless, the creatures, for whose welfare the owner is confessedly responsible, be better treated at a diminished outlay.

When a dumb slave fails in the service of some affluent proprietor, all that might be done is not accomplished when the assistance of a veterinarian is sought. It is not sufficient that drugs are paid for, and that the doctor is in constant attendance, or that the building, in its arrangements and details, attests to the vigilance of his measures. No: a man is bound to investigate, to judge conscientiously, and to determine those means to be adopted best calculated to promote the health and happiness of the inmates of his stable. Had this duty been discharged, many processes, still sanctioned by custom, might have fallen into disuse ; some habits, now indulged, might have been discarded ; while a few objectionable measures might have been altogether forbidden, as useless formalities, and needless cruelties.

Horse balls—particular forms of veterinary medicine—are

generally sent to stables by the dozen. Physic is thus placed at the pleasure or the caprice of ignorance to administer.

The aloetic mass is the common purgative of the stable. So general was the use of aloes, and so unquestioning appears to have formerly been the confidence lavished upon its operation, that this medicine always took precedence in every sickness, and ultimately engrossed to itself the significant term "physic." "Prepare this horse for physic," signifies, in stable phraseology, first, give such an animal four or five bran mashes during 24 hours, and, at the expiration of this time, administer to the said creature a dose of aloes. The groom cannot imagine a medicinal compound to be worthy of the title "physic" unless it be capable of producing most violent drastic effects. Other medicine will move the bowels of the horse, and bran mashes given three or four times daily will relax the intestines more gently; but the stableman is never satisfied without aggravating the result with a dose of aloes.

The horse's body does not quickly respond to opening medicine, but the action, once elicited, is not invariably easy to command. The animal's life is frequently a prey to a potent purgative. The veterinarian knows that the different creatures vary much in their capability of swallowing amounts of aloes; that the dose which will not move one quadruped may destroy the inhabitant of the next stall. One creature will imbibe two ounces of the drug without marked effect; another will be shaken by the action of less than half an ounce of the preparation. Nevertheless, the stable-man always craves for aloes, and always experiences an odd delight when watching for its hydrogogue operation.

In the majority of cases bran mashes, in the place of aloes, could be advantageously supplied, which are readily made as follows :—Put a peck of bran into a perfectly clean stable-pail. One person should stir the bran as briskly as possible, while another person, with speed, empties a sufficiency of water

just below boiling point into the pail to render the contents
a pultaceous mass. The vessel is then covered up, and when
it has become cool, the pudding is thrown into the manger.

However, one horse will devour bran mashes with avidity;
another will not touch them. This animal will not partake
of the potion, unless it be partially warm; another will not
eat until it is perfectly cold; while most will partake of the
mess if it be flavoured by the admixture of a little salt, or a
few crushed oats.

So it is, also, with water. Certain horses, when feeding
upon bran mashes, refuse all drink; others enjoy frequent
draughts of cold fluid, while a third set seem to crave for warm
water; and a fourth will neither imbibe freely nor entirely
abstain, being wholly indifferent as to the temperature of the
liquid. Thus, the order which is inserted in most books, to
give to the horse, after the animal has swallowed a dose of
aloes, copious draughts of warm water, is frequently rendered
futile; for, as the proverb teaches,—" One man may lend the
horse to the pond, but forty men cannot make the quadruped
drink."

Bran mashes, however, will act without the aid of repeated
doses of warm fluid. Of themselves they do not debilitate,
though from the length and size of the horse's intestines,
purgation cannot be long maintained without inducing serious
exhaustion; and it is never safe to work the animal while any
looseness is observable. A tendency to inflammation is often
announced by repeated and liquid discharges; therefore, never
let the horse be taken out while the bowels are in a state of
excitement, for exercise may increase that action to one of
positive disease. Bran mashes, however, are the safest and
the gentlest of laxatives. Any condition may be induced,
according to the number and frequency of the potions. In
general, they act mildly, without inducing that bodily discomfort or that constitutional weakness which throws the

animal out of condition, and renders rest an absolute necessity
for recovery. Altogether, these mixtures are the best and the
safest laxative of the stable ; but even these should never be
administered to the horse without the special direction of the
proprietor.

On the other hand, aloes, in no form, can be administered
to some horses. Very many cannot receive a full dose of the
drug. Several can only with safety swallow the medicine
when highly spiced, or in solution ; while a few are all but
insensible to the action of the agent. Alarming spasms often
follow the exhibition of a moderate quantity of aloes, which
always renders the quadruped sickly ere the effects are visible.
The drug, in most instances, lies dormant 24 hours, during
which period the appetite is lost, the spirits oppressed, the
coat dull, and the entire system evidently shaken. It is not
esteemed prudent to work the patient till several days' rest
have been allowed for its restoration.

It used once to be the custom to trot the animal which was
sickening under a dose of aloes ; but experience has shown the
danger of the habit. The horse is now left in the stable, has
an extra rug thrown upon the back, while a pail of warm
water is in most instances placed in the manger. Where safe,
it is obviously unnecessary to ride the quadruped which is
sickening under aloes, since the loss of appetite shows the
medicine has affected the system, and the natural effects of
the physic may, therefore, be anticipated.

Very many animals, when suffering from chronic debility,
may be killed by a moderate dose of aloes, while many never
sufficiently recover from purgation to do a day's work after
the medicine has ceased to operate. Of all the preparations
the veterinarian has at his command, we do not know one
which exerts so decided an effect upon the constitution ; nor
does the veterinary pharmacopœia contain an agent which
demands more care and caution in its administration. Several

potent caustics rank among the most common of horse medicines, and should never be given, excepting under the direction
of a duly qualified veterinarian ; for, in the first instance, some
grooms are unable to administer medicine satisfactorily, either
in the form of ball or otherwise ; and, secondly, even allowing
that some grooms are moderate manipulators, they cannot
have the experience of the veterinarian or his running doctor,
—persons who are daily giving medicines to animals in their
various forms.

There are some horses of a savage nature, and who are
constantly prone to bite, but more particularly so during the
exhibition of a ball. To obviate the accidents likely to result
from this vice, and to assist the operator during his manipula-
tions, an instrument called a " balling-iron " has been invented.
A " balling-iron is simply a piece of metal, so shaped that
when thrust violently between the creature's jaws, it forcibly
holds the mouth open. Therefore it will certainly prevent
biting ; but an irritable or a fearful horse can rear up and
strike with its fore feet. Such an animal is not entirely subdued when the iron is adjusted. The man using a balling-iron
has, therefore to guard himself from blows rapidly dealt with
the fore hoofs of a desperate animal.

He has also to be ready at the slightest intimation of an
intention to rear, so that he may withdraw his arm on the
instant, otherwise the operator is dragged upward with the
elevated crest, and, hanging by the inserted member, he is
very lucky if a broken limb does not reward his tardiness.
The use of the balling-iron, consequently, is not free from
danger ; and, in practice, it will be found safer to subdue by
kindness than to partially conquer by the employment of mechanical restraints.

The most common form of balling-iron is constructed according to the model indicated in the following illustration.
The circular piece of metal is inserted into the mouth of the

animal. A straight bar is attached to either side of the metallic ring, the design of these last being to steady the instrument after it has been forced into its proper position. Through the circle the operator's arm is thrust, and the iron ring affords security so far, as it disables the jaws from closing upon the member. But, though safe in one direction, such a protection also creates its particular peril ; for, should the horse rear, the

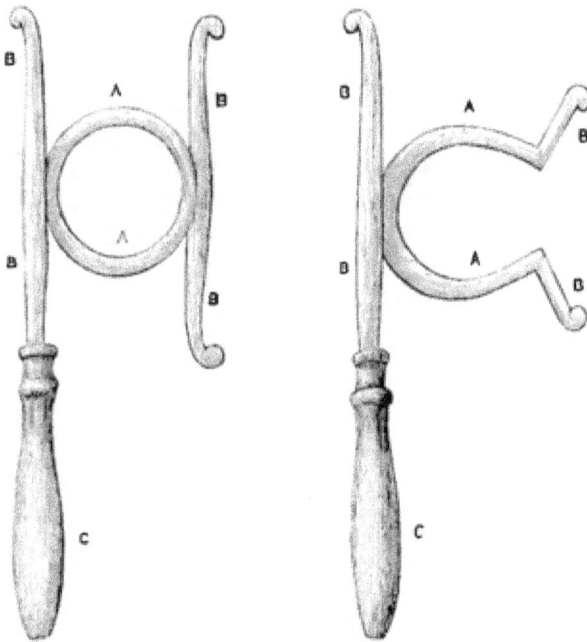

THE COMMONEST FORM OF BALLING-IRON.

A A. The ring of iron, which being forced into the animal's mouth, keeps the jaws asunder.

BB, BB. That portion of the metal which steadies the ring by remaining against the jaws.

C. The handle.

THE IMPROVED FORM OF THE COMMON BALLING-IRON, WHICH AFFORDS A PROBABILITY OF ESCAPE FOR THE OPERATOR'S ARM.

A A. The part forced into the mouth.

BB, BB. The part which remains against the jaws.

C. The handle.

arm, being surrounded by a metallic rim, could not be withdrawn with the speed requisite to insure the operator's safety. The suspension of the man's body is almost certain to provoke the fracture of his imprisoned limb ; consequently, to remedy that evil, the improvement, indicated by the right-hand illustration, was introduced.

The circle in the above was left free on one side ; thus the inexpert had a little more time allowed for their movements. The arm could be retracted with greater ease, and the former danger was, in a great measure, removed. Still, this new shape was not wholly satisfactory. The form was fixed : horses are not all of one height, one breadth, or of one capacity. There are small creatures designated ponies, while horses are not rarely encountered of enormous proportions. As the iron had no power of being adapted, the form that should prove not large enough for one, might be altogether disproportioned to another quadruped.

The weight of metal necessarily employed to assure the requisite strength, also rendered it inconvenient for a veterinary surgeon to carry more than one of these bulky articles ; and though small was the amount of ingenuity which had hitherto been lavished on the improvement of the thing, for years it continued of the last character. Mr. Varnell, assistant professor at the Royal Veterinary College,

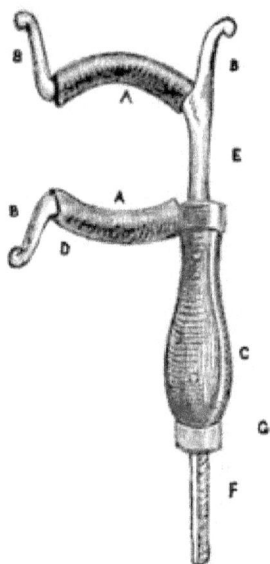

A NEW BALLING-IRON, INVENTED BY PROFESSOR VARNELL, OF THE ROYAL VETERINARY COLLEGE, LONDON.

A A. India-rubber tubing to protect the mouth from the harshness of the metal bars.

B B B. Side pieces to keep the iron in its situation.

C. The handle.

D. The lower bar, attached to the handle.

E. The side piece, which can be raised or depressed.

F. The screw, at the extremity of the side piece.

G. The nut which, fastened to the handle, acts upon the screw, and fixes its position.

however, appears to have entirely removed all former objections, and to have invented a balling-iron, which seems to
possess all the qualities that such an instrument is capable of
exhibiting. The restraining bars of this last amendment are
formed of polished steel, and are covered with a stout piece
of India-rubber tubing; thus, in some measure protecting
the mouth of the creature from injury by what hitherto was
the exposed metal. The lower bar, moreover, is attached
to the handle, and the handle can be readily raised or depressed by turning the nut situated at its base. It can, therefore, be quickly adapted to any possible capacity of jaw.

Such a form of immunity is, however, seldom sought, save
by the very inexperienced in the veterinary practice. A few
years of active employment enables any person to discard this
defence. A sufficient security is in all ordinary cases afforded

THE USUAL MANNER OF GIVING A BALL.

by the horse's tongue, which, when a ball is about to be ad-

ministered, is grasped by the left hand, and withdrawn to the
right side of the mouth. The hand thus employed is fixed,
being lightly pressed against the inferior margin of the lower
jaw; for, when retained in such a position, the tongue is
pressed upon the foremost of the huge molar teeth. Of course
the animal, thus held, cannot approximate its jaws so as
seriously to harm the operator without biting its own flesh;
by that circumstance is safety supposed to be rendered certain.
But should violence be exerted, animal fear is apt to be
superior to bodily pain; the tongue and arm may be simultaneously bitten through. The practised veterinary surgeon,
however, takes advantage of the first emotion of surprise
which the creature experiences at the liberties taken with, and
the indignities offered to, its person. Having the ball ready
in the right hand, he, standing on the left side, quickly intro-
duces the bolus into the wondering quadruped's mouth.

The medicine is lodged at the back portion of the tongue,
whence, as the horse does not expectorate, the creature has no
ability of expelling it, save only by coughing. During the
spasm, which accompanies this act, the soft palate is raised,
and the ball is carried outward with the volume of violently-
expired breath. Some horses acquire a habit of thus return-
ing all forms of physic, and will cough up a ball twenty times.
Such a circumstance illustrates the necessity of distracting the
attention of the quadruped, the instant the hand is retracted;
for, in the confusion of the moment, the most inveterate
"dodger" may be surprised into swallowing any abhorrent
morsel.

The hand, during the delivery of the ball, being rapidly
thrust into the mouth, is frequently cut by the sharp edges of
the molar teeth. No knowledge, which has hitherto been
attained by veterinary science, can point out the animal
possessed of grinders of this dangerous description, and the
only protection as yet suggested is to cover the hand with a

glove. But a glove cannot be washed and dried so readily as the hand; it, moreover, is highly objectionable to introduce the saliva of one animal into the mouth of another, as disease

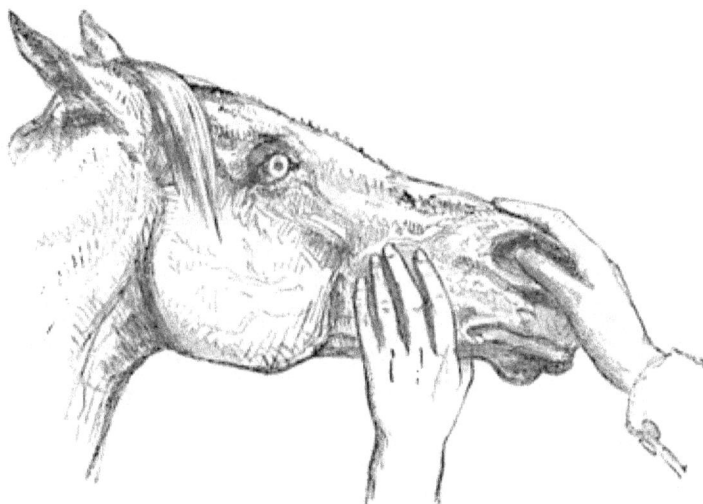

THE CUSTOMARY MODE OF DISTRACTING THE HORSE'S ATTENTION, AFTER IT HAS RECEIVED
A BALL.

may be thus conveyed from horse to horse; also, it being impossible to provide a new glove with every fresh patient, the protection is not universally adopted.

The medicine being delivered, the hand is quickly withdrawn, and the jaws of the animal are clapped together. The nose is then rubbed somewhat roughly, for—the upper lip being the organ of prehension, as well as the seat of feeling, in the horse—this part is excited, with the design of preventing the quadruped from dwelling too intently on the unpleasant nature of the substance which has just been forced into its mouth.

The best mode of holding a ball during administration is to place the ball between the fore and middle fingers, and to keep the hand level with them; in this position it is introduced, and, when it has been advanced to the desired situation, a separation of the fingers allows the morsel to drop into its place. After the exhibition of the above physic, the horse's attention may be engaged by applying gentle friction to his throat, until the bolus has been swallowed.

MANNER OF DELIVERING A BALL.

The delivery of a ball is, however, rendered difficult in proportion to the number of persons who surround the animal, and to the noise generally made on such occasions. The experienced veterinarian goes alone to the head of his patient,

THE QUIET METHOD OF GIVING A BALL.

and plays for a time with the quadruped's face: confidence

being thus established, the practitioner gently withdraws the creature's tongue. This being accomplished, of course the jaws are sundered, when, without any sign of flurry, the hand is introduced into the cavity, and the medicine properly lodged. After such a manner the practised veterinarian gives many balls in the course of the day, and is hardly ever known to fail. Indeed, were the practitioner, when going his rounds, to wait till four or five assistants could be collected ere he administered the requisite medicine, the duties of the day could never be discharged.

The physic being introduced into the mouth, the person who has undertaken to deliver it should on no account esteem his business finished, and thereon leave the stable. He should proceed to the left side of the horse, and watch the neck. In that position, when the animal swallows, any substance can be seen to travel down the gullet : this proof having been witnessed, the building may be quitted with a safe conscience.

BALL PASSING DOWN HORSE'S GULLET.

Drinks or draughts constitute another usual mode for administering medicines, and are considered by most prac-

titioners to constitute both an easy and effective medium for
exhibition. A drink is usually given through the mouth, but,
in some cases, such as of locked-jaw, *i.e.*, when the mouth
cannot be opened, medicine in solution is caused to find its
way to the stomach through the medium of the nostrils. At all
times administration by this channel is fraught with danger;

FIG. 1.

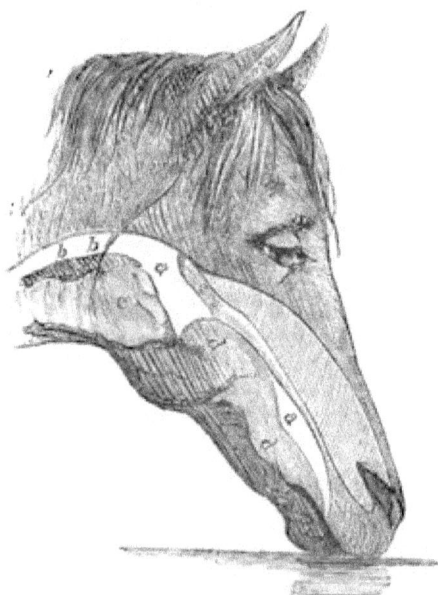

DIAGRAM (FIG. 1), EXPLANATORY OF THE COMPOUND ACT OF DRINKING IN THE HORSE.

a a. The water drawn into the mouth, and forced into the fauces by the compression of the
forward part of the tongue, and the enlargement of the backward portion of the organ.

b b. The fluid passing down the œsophagus or gullet.

c. The larynx, lowered to admit the passage of the liquid.

d d. The tongue, dilated at one place and contracted at another.

e. The soft palate, floated upward and effectually closing the nasal passages;

for the liquid, sometimes charged with caustic contents, may
flow to the lungs, and thus produce death from suffocation.
The natural receptacle for food and fluids is the mouth, and
any attempt to administer them through any other channel is

an unnatural action. To render this subject more intelligible, the process which enables the horse to imbibe liquids shall be here detailed.

The mouth of this animal is peculiar for having at its backward extremity a fleshy screen, which hangs pendulous from the bony roof. This soft palate explains why the quadruped, under ordinary circumstances breathes only through the nose, and why, when it vomits, the regurgitated matter is ejected by the nostrils. That speciality is of service, however, during the act of imbibition. The posterior entrance to the nasal chamber being open, and the head in a pendulous position, were there no special provision to the contrary, the water, after having passed the mouth, would, from the mere force of gravity, have a tendency to return by the nostrils. This actually occurs whenever cold, strangles, influenza, sore throat, &c., interferes with the activity or the health of these parts now under consideration. Disease renders the organ sensitive, and tenderness makes the animal exert its volition to prevent the employment of the inflamed structure. In consequence of this cause, the nasal chambers are imperfectly closed, and a great portion of the fluid imbibed by the mouth flows forth again through the nostrils. Such a tendency to gravitate is, during health, effectually prevented by the soft palate. Before any substance can pass from the mouth towards the throat, that appendage must be raised, and its rising closes the posterior entrance to the nasal chambers.

The tongue is the primary agent employed when the animal slakes its thirst. The backward portion of the organ is contracted, and the forward part compressed by muscular volition (d d, fig. 1). A vacuum would thereby be created, were not the water propelled by atmospheric pressure into the void thus formed (a, fig. 1). The posterior of the tongue is then relaxed, while the anterior division of the organ is pressed against the roof of the mouth (d d, fig. 2). The fluid is thereby

4 *

driven to the backward part of the cavity (*a*, fig. 2). The tongue, during the act, continues to alternate the states of contraction and relaxation, each motion of the lingual agent serving to pump the water into the fauces (*a*, fig. 1). But,

Fig. 2.

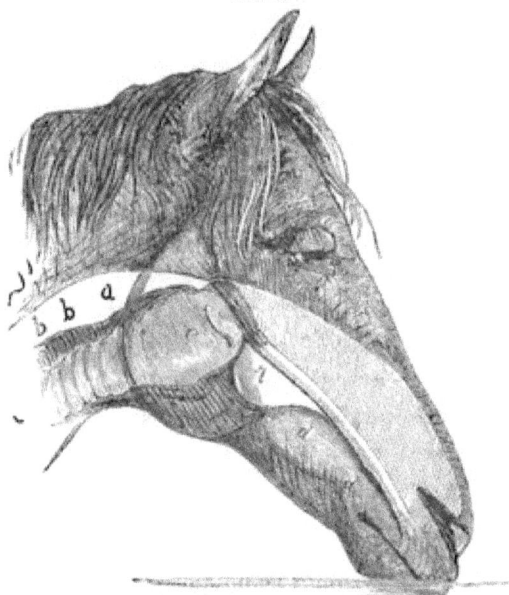

DIAGRAM (FIG. 2), EXPLANATORY OF THE COMPOUND ACT OF DRINKING.

a a. The water driven backward by the forward dilatation of the tongue and the upward movement of the larynx.

b b. The full current forced down the gullet.

c. The larynx propelled against the soft palate.

d d. The tongue dilated anteriorly, and compressed posteriorly.

e. The soft palate.

before that can be accomplished, the soft palate must be elevated. The soft palate (*e*, fig. 2) then closes the nostrils (*e*, fig. 1); and also in its course to take this position sets in motion the cartilages of the larynx. The last cover over and effectually protect the windpipe (*c*, fig. 1), and the fluid, forced onward by the contraction of the tongue, passes into a secure chamber, the roof and floor of which are but of temporary formation (*a*, fig. 1). Here it remains only during the inactivity of the larynx. The upward motion of the latter body (*c*, fig. 2) propels the fluid into the pharynx, whence involuntary contractility sends it into the gullet, the muscular action of which tube conveys it onward to the stomach (*a b b*, fig. 2).

From the foregoing explanation, the reader is in a position to judge whether the nasal chamber is a fit passage for acrimonious mixtures, since he now understands the evident pains Nature has bestowed to prevent the temperate fluid, of which the animal customarily partakes, from intruding upon the elaborate, delicate, and highly sensitive membrane that lines the air passages.

The usual mode of giving a drink is, moreover, a complex business. A twitch is mostly kept in regularly-appointed stables, and the string or loop is fixed over the animal's upper jaw, prior to other measures being proceeded with. The groom then grasps the stick, and takes his place by the shoulder of the horse. At a previously arranged signal, he raises the pole; the string, paining the gums, under which it is fixed, causes the head of the quadruped to be elevated, when the fluid contained in a bottle (to be afterwards described) is next carried upward, two fingers of the operator's left hand being fastened on to the gums, so as to further expand the jaws, and enable the veterinary surgeon to steady his body while straining to administer the medicine.

In administering a drink the operator should stand on the right side of the animal, whilst the attendant at the twitch should take his position on the left side. Years ago drenching horns and tin bottles were used as media for the exhibition of mixtures; but of late years they have ceased to be employed, and their place has been filled by a strong glass bottle capable of holding the whole of the mixture to be given. In this case the operator is not compelled, as was the case with

the drenching-horn, to be continually filling the receptacle
when empty, which caused delay, and in great measure frus-
trated the ease of administration. But from the foregoing it
must not be imagined that a drink should be poured down the
animal's throat all at once. It must be given in small quantities at a time ; a quart should be divided into 15 mouthfuls,
and no attempts should be made to administer a second
mouthful before the former one has been swallowed, and for
the simple reason that, should the operator fill the mouth too
full, or the animal cough, the administrator is likely to be

GENERAL METHOD OF ADMINISTERING A DRINK TO A HORSE.

saturated with medicine, as any irritation occurring to the
larynx is invariably productive of this effect : consequently

any attempt to give drinks otherwise than gradually, is productive of waste, and necessitates the taking of only a portion of the intended draught, thereby frustrating the object in view, namely, the exhibition of the whole of the prescribed dose.

To some horses, viz., those of a quiet disposition, drinks can be administered without the assistance of a twitch, and doubtless the " quiet method " of giving a drink recommends itself to our notice before any other. It consists in the operator approaching fearlessly the animal in fondling, and pro-

GIVING A DRINK, ACCORDING TO THE QUIET METHOD.

ceeding to such trivial familiarities as may establish perfect trustfulness between man and his dependant. So soon as the steed's confidence is gained, the animal is all submission to the pleasure of its superior. Then let the practitioner uncork the bottle, and, putting the left hand gently under the quadruped's jaw, empty with the other the contents, gradually, through the interspace which divides the incisors from the molar teeth.

But, when adopting the above plan, the operator must be alone. No noisy or officious assistant must be near at hand to excite alarm or to create distrust. No pain must be inflicted; no angry words should be employed; no violent or hasty action ought to be used to frighten native susceptibility. All must be quiet: should the animal be slow to swallow a nauseous draught, the creature must not be scolded for a natural dislike, but it should be encouraged by kind and cheerful accents, spoken as softly as though the words were addressed to a sick child. So alive is the equine heart to the seductiveness of benevolence, so unsuspecting is the full confidence of its species, and so happy is its spirit made by the praises of its superior, that rather than not deserve his commendation, it will gulp down the most distasteful solution.

BLISTERING AND FIRING.

The system of blistering horses (which consists in applying to some part of the animal's body, but more particularly the legs, a compound composed of cantharides—Spanish flies,—and lard, and its potency is sometimes increased by the addition of turpentine), has been practised for many years in this country, and no doubt in certain cases with marked benefit, provided that only one or two legs at most have been blistered at one and the same time.

Horses have perished under the agony attendant upon the blistering of all four feet. It is, however, still a recognized custom for horse doctors to score a leg, or sometimes two legs, with the red-hot iron, and, over the lines thus created on a living frame, to apply a blister. To fully appreciate the abhorrent barbarity or the inutility of such a custom, the reader must recognize that animals suffer awfully from the wounds occasioned by fire, and understand that the sores are newly made when the irritating compound is placed upon the tender parts. A blister necessitates that the grease, which contains the extract of the fly, should be thoroughly rubbed in; therefore the horse, when blistered, after having been fired, has to endure the friction of a rough hand, applied with all the coarse energy of an uneducated man, made upon a member, smarting under the agony produced by the agent of which the creature has an instinctive dread.

Blisters, as at present used, are far too powerful. Were they diluted with three times their bulk of bland oil, or of solution of soap, they would be equally effective, and far less dangerous; but, unfortunately, there is a prejudice among the partially educated, to which class nearly all veterinary surgeons belong, in favour of potency in their applications. Such persons seem to reckon the benefit to be produced according to the strength of the agent employed. By what other reason is it possible to explain the foolish perversity which still clings to the abuse of the heated iron? By what other motive can we account for the prejudice which tempts the use of the fearful blistering oil, as now commonly exhibited?

The parts of the horse most generally blistered are the legs, and the explanation commonly given to excuse the folly, is a desire "to freshen the old animal on the pins," or "to brighten up the manner of going." The legs are parts of the living frame, and one part can hardly fail without the system sympathizing. The legs may be the means of progression, but it

is the life which puts them in action, and it is the nerves
which transmit energy to the muscles.

There is a maxim, first made known by John Hunter,
and subsequently recognized by the profession of which
he was the ornament. This maxim declares that "two great
inflammations cannot exist in the same body at the same
time." Upon the truth of this discovery the practice of
counter-irritation is based. Then to fire and to blister simultaneously may increase the torture of the poor existence thus
barbarously treated; but, according to the doctrine, largely
accepted by the medical profession of this country, the double
process accomplishes nothing surgical or curative, since the
blister must destroy the action of the fire; and the man who
is greedy to obtain the benefits of both operations, secures the
advantages of neither measure.

To blister, however, is a very antique custom; so, also, is
the application of fire, which was first performed upon the
human body. Old medicine does not bear a very good character, and only exemplifies the much which suffering can
endure, or the little which cruelty can accomplish. The blister
is, according to present veterinary practice, employed more
often to gratify the passing whim of some wayward proprietor,
than with any medical intention, or with the remotest regard
for the quadruped. A man, while lounging through the stable
of an evening, without the slightest pretence to medical knowledge, may conceive he will have the entire stable blistered
"right through," and few veterinary surgeons will presume
to expostulate with so wild a notion.

The compliance of the professional attendant is often, however, in strict keeping with opinions implied by the expression
commonly employed by "horsemen." Thus it is very general
to hear these persons speak of—"a good horse with battered
legs"; "a beautiful animal, but with legs that have done
their work"; "an excellent frame, but not having a leg to

stand upon," &c., &c. Such phrases are sheer nonsense, but
they serve to countenance the equine superstition which
regards the legs as distinct from the body. The stable-man
cannot conceive a want of liveliness in the motions to be one
of the indications of failing health; yet this symptom pervades
all nature. It is exhibited by beasts, by birds, by fishes, and
by insects; nay, the very vegetables, when disease attacks
them, no longer spread their branches to the breeze, but droop
their heads, and incline their bodies earthward.

To propagate such opinions, however, must destroy much
of the power so dearly loved by the horse-owner, and abolish
much of the pleasure such a person experiences when survey-
ing his long rows of miserable dependants! It is astonishing
how unfitted human frailty is to possess absolute authority in

BLISTERING A STABLE-FULL OF OMNIBUS HORSES.

any shape! The men who live and think in stables are never
so happy as when exercising their despotic power. The above

illustration is an example of this fact. An omnibus proprietor
has entered to speak with a veterinary surgeon, who is witness-
ing the man's orders fulfilled on the forelegs of a wretched
stud. Let the reader contemplate this engraving, and he will
soon perceive the animals stand in need of something far less
costly than any mixture which can proceed from the cheapest
pharmacy.

It will be remarked that the creatures represented above
are separated by "bales," or by long poles, suspended by
chains from the ceiling. This kind of arrangement permits
more horses to be packed into a limited apartment, and is,
therefore, adopted whenever the expense of lodging becomes
a primary consideration. It will also have occurred to the
spectator that the roof is depicted as very low, and the gangway, or free thoroughfare, behind the animals, is exhibited as
exceedingly narrow.

Now, creatures imprisoned in such a building, are actually
perishing of starvation! The food, the water, and the medical
attendance, may each of its kind be unexceptionable; but the
animals housed in such a locality soon droop from positive
inanition. To breathe, is the primary necessity of existence.
There is no living thing that can thrive where air is excluded.
The quadrupeds, represented above, have to pass 22 out of
every 24 hours in a locality barely lofty enough for each to
stand upright in. Let the reader, knowing the duration of
captivity, conjecture how long it will be ere the huge lungs of
a horse have inhaled and contaminated the limited amount of
atmosphere which the place can contain, even where such an
abode contemplated as the dwelling of a single subject.

It is true, such sheds are seldom air-tight. Were all
draughts excluded, the prisoners would speedily be released
from their captivity; but the wind-holes, though large enough
to prolong misery, are too small to render such places the
abodes of health. The wretched inmates cannot be tortured

into a show of activity. When will the Legislature, in its wisdom, notice these hot-beds of contagion? When will it empower the police officer to enter any stable, and authorize him to destroy the animals therein, hopelessly diseased, and purposely concealed? Who can, viewing the stables where the hardest worked of the equine race are stowed away, wonder that glanders is rarely absent from such nurseries for contamination?

It is a common, but most mistaken custom, to purge a horse before blistering his legs. The intention is to remove any lurking irritability out of the animal's system; but such irritability will most probably be provoked by their coarse and potent blistering agents; therefore, a purgative, by increasing the debility, is only likely to render the quadruped more sensitive to outward impressions. A nice "freshener" is embodied to the eye of reason, in a drastic purgative followed by an active irritant, applied to a most sensitive part of the body!

Whenever a blister is adopted, the action rather depends on the amount of friction which accompanies the agent, than on the quantity of the vesicatory that may be employed. The friction should be regulated by the condition of the surface on which the oil has to act, and all adjacent tender places, as the points of flexion in the joints,—parts where the skin is thin or is thrown into crevices, should be previously covered with a layer of simple cerate, after the method exemplified in the next illustration, wherein the back of the pastern is exhibited as thus protected.

The part to be blistered must be covered with a small quantity of the ointment just sufficient to adhere firmly to the hair and skin; for, if too large a quantity be used, the blistering agent has a tendency to run, and thus invade adjacent parts, or those not intended to be subjected to its action.

It is necessary, after applying a blister to the legs of a horse, to fix his head to the rack for 36 hours, in order to prevent him from gnawing his injured legs, after the expiration of which time the blistered surfaces should be well washed with soft soap and warm water, and the following day, *i.e.*, when the legs have become dry, and encrusted with a solid

THE BACK OF THE PASTERN AND THE HEEL PROTECTED BY BEING COATED WITH A
THICK LAYER OF SIMPLE CERATE.

exudation, it is humane to soften the part with some emollient liquid, viz., lead liniment, which is made by mingling one part of Goulard's lotion with two parts of olive oil, whereby a thick creamy compound is formed : the grease soothes the harshness of the exudation, while the lead serves to mitigate any pain which may reside in the part. This mixture, being well shaken, is applied to the surface by means of a soft sponge or brush.

The "liniment" usually causes the "crusts" to fall off; but the hair generally comes off at the same time, testifying the severe irritation to which the skin has been subjected.

The most pliant medical individual—the pedantic man, who always acknowledges everything emanating from the schools to be correct, would, we imagine, be puzzled to discover any necessary connection between the processes of *balling*,

blistering, firing, and *bleeding;* yet, somehow, the four operations are associated in veterinary practice. A ball reduces
the bodily activity; a bleeding lowers the action of vitality;
irritants are thought to stimulate organs to which they are
applied, but to lessen the general tonicity. An animal,
subjected to the first action, appears fitted to dispense with
the second, while the last two seem somewhat similar to

HORSE'S LEG AFTER THE APPLICATION OF A BLISTER.

the first. But there is no accounting for incongruities when
men, deserting reason, consent to adopt routine as a guide in
the treatment of so capricious a development as disease.

To lose blood was once deemed a healthful custom by the
human race. Then, horses were regularly depleted every rise
and fall. An old practitioner can remember the period when,
on a Sunday morning, he beheld long sheds full of agricultural
quadrupeds waiting to be bled. The fleam used to be struck
into the first horse, then the entire row were, in succession,
similarly treated. The operator afterwards returned, and,
pinning up the wound which had been made in the neck of

the first animal, again moved down the line, pinning as he
went. No account was taken of the amount lost by each
patient, nor were any pains thought needful to control the
current that flowed upon the ground; but the creatures did

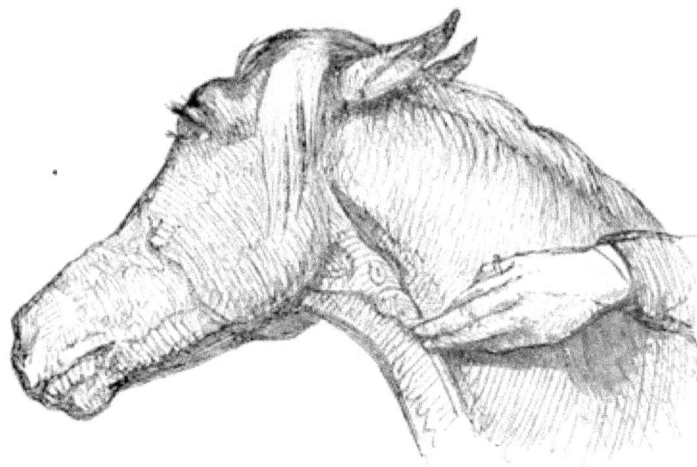

RAISING THE JUGULAR VEIN.

not all suffer an equal depletion. The fleam was soon struck;
to pin up, however, took a comparatively long time for its
performance. The first horse of the group, therefore, lost but
little blood, while the last of the line bled for a considerable
period before its turn to be attended to arrived.

The foregoing anecdote will show how nice our ancestors
were in their operations; but it is sad when we reflect that all
this carnage was a sacrifice made to a mistaken idea. Human
medicine has abandoned the antiquated custom. Veterinary
physic, however, is not quite so versatile; still many quiet
spots in the country may be found, where old physic is in
force, both with the employers and the practitioners. Dogs,
even in the metropolis, are sometimes bled, and there still

exist persons who esteem the use of the lancet upon these animals to be a laudatory accomplishment.

No matter what may be the age, the condition, or the occupation of the horse, certain practitioners always discover that the mute drudge requires depletion; thus an unscrupulous man may at most times earn a ready shilling by performing an easy operation. The strangest fact is, that most rural proprietors love to see the purple life drained from the necks of their possessions; and bitter are the reproaches usually lavished on the veterinarian, should a horse perish of any disease without the fatal termination having been hastened by the favourite measure. Indeed, so fully are several country practitioners aware of this probability, that it is customary with them, when alone, to strike the vein and to pin up the orifice immediately. The necessary sign can then be adduced, should death end the case, and a professional reputation be thereby saved from the assaults of aggravated stupidity.

To show how necessary the adoption of venesection is considered to be by many persons, Mr. Mayhew states, that some years ago, a medical man, then residing in Westbourne-terrace, kept a well-stocked stable. The family going out of town during the autumn, some of the animals were allowed a few weeks' "run at grass."

When the horses were taken up, none were found to have been benefited, but one was discovered to be much worse for its period of liberty. It was very weak, and its constitution evidently was shaken, for nothing seemed capable of invigorating it. If put into harness and driven merely round to the street-door, the body was sure to be white with perspiration, and the poor quadruped exhibited signs of exhaustion. If permitted to remain in the stable, the creature would generally be found with the head depressed, the corn untouched, the breathing audible, and the body leaning for support against the trevise.

5

The animal was in this state when the family again left London for a few weeks : the horse was taken with them by railroad. Before they quitted town, Mr. Mayhew found occasion to speak with the proprietor, and informed him that, during the sojourn of the family in the country, it was probable the urgency of the symptoms would necessitate the calling in of a local veterinary surgeon; therefore, the proprietor was warned that the ailing quadruped was on no account to be bled, for, to deplete a life in so exhausted a condition was positive slaughter.

This event literally happened. The symptoms suddenly became alarming : the attendance of the nearest veterinarian was requested. To him the warning given to the proprietor was repeated. The gentleman replied that his adviser had not seen the animal in its then serious state, or he could not have tendered such advice. Medical etiquette forbade positive injunctions. The operation was performed, and the family returned to town leaving a carcass behind them !

It is very seldom that the system of a horse, when doing full work, can endure depletion. The labour is exhausting, and the toil is sufficiently severe to employ it all, had the animal twice its normal energy. Many observant stable-men are of opinion, that, nurture as they may, the provender consumed cannot be equal to the work. There are, however, too many persons who study to under-feed, and who, nevertheless, are morally convinced that every quadruped in their keeping not only possesses a sufficiency of vigour, but can part with a gallon or two of blood twice in every year, with positive advantage.

Here are two opposite convictions, and the cost of horse-flesh to each party, could we inspect the private accounts, would certainly best settle the dispute. But as men mostly object to laying open their books to public investigation, we must, therefore, endeavour to decide this point by drawing inferences, after having submitted the lives of most quadrupeds to review. None, except the wealthy, keep horses, save for use. The feelings of men are seldom gratified by feeding idle animals. Two horses very commonly have to perform extra duty, while the master is looking about him, and in no haste to purchase a third labourer. Rarely do we find three animals are kept where the owner has full employment only for two of his slaves.

The horse, therefore, is generally worked to the limits of its strength. That there may be no doubt upon this matter, the person who has to judge of its capabilities is he who has an interest in the amount of an animal's exertions. The fact is, however, proved by the wonder excited when a quadruped is recorded to have reached the natural period of its existence. The great majority of horses in this country perish of exhaustion before their maturity has been attained. The sad reality that, of the numbers reared in England, the great majority of humanity's humble, obedient, and willing slaves are goaded to early graves before all their second teeth are up, and before the consolidation of their bones fits them to endure the strain of fatigue, too fearfully establishes the fate which beauty and submission receive at the hands of avarice.

There is something wrong in the creature who can thus abuse all that serve him. Had the horse twenty times its present strength, it would still be below the point of human requirement. It is a very painful occupation to look into a London street, and, having an understanding which can interpret equine significances, to observe the lame, the deformed, the starved, the over-loaded, and the weary animals, staggering along the thoroughfare, but to perceive none without the goad, to enforce exertion, flourished by its side. Yet the creatures, thus used, unconscious of a holiday, and worked through sickness or through suffering, are thought, by some persons, to possess such a redundancy of health that they can support

5 *

or be benefited by the life's blood being drained, at stated periods, out of their wretched bodies!

Nevertheless, it is possible a timely depletion may, upon certain occasions, save life, such as in cases of inflammation of the serous membranes. Neither the present reader nor the writer may witness so rare an occurrence; yet, because of the possibility, every horseman should be equal to such an emergency. For the performance of so trivial an operation certain tools are imperatively necessary. The first among these is a blood-can, or a tin pail, which is generally divided, by indented lines, into eight equal sections. The receptacle being made to contain two gallons, each compartment, when filled, indicates a quart to have been withdrawn. Wretched horses have been drained to a greater extent even than two gallons; but, should the reader possess a blood-can, it is hoped that it will be indeed an extreme case in which he would behold the vessel once filled.

A BLOOD-CAN, WHICH IS MARKED TO INDICATE WHEN A QUART OF LIQUID HAS BEEN EXTRACTED.

One or two quarts should be the limits of an ordinary venesection; but even that quantity may be of much more service when aiding the circulation than when withdrawn, and permitted to coagulate apart from the body. Many practitioners, however, deplete without either excuse or justification. Having opened a vessel, they will allow the stream to flow until the poor horse staggers. Some are proud not to possess a blood-can; but they hold up the stable-pail to catch the vital current, and are quite content that the most ample drain of the system, conducted under their supervision, cannot be otherwise than restorative.

The next instrument requisite is a fleam. This article is

much preferable to the lancet, though there exists a species of foppery among veterinary surgeons which tempts them to employ human implements. For that reason they flourish a

FLEAMS, OF THE NEWEST FORM, OPEN AND SHUT.

lancet as the more scientific indicator. A lancet is, certainly, necessary to puncture the eye vein, which is visible upon the cheek of the horse; but, as regards a vessel which is as large as a cart-rope, for such is the dimension of the animal's jugular, this last cannot demand the exhibition of vast scientific attainment to pierce it, or admit of the display of nice manipulation in him who operates on such a structure. For this reason the old-fashioned fleam is very much to be preferred. Assuredly it does not appear so pretty as the lancet, but it always cuts with certainty, and leaves a limited orifice, whereas the more genteel blade has inflicted awkward gashes upon living flesh when the creature proved restless under its infliction.

The instrument with which the veterinarian extracts blood has been represented, having the blade bared, and having it closed. It is readily admitted not to be of an inviting aspect, but it is not in reality quite so barbarous as it appears to the beholder. The point which projects from one side of the blade marks the extent of its cutting surface, and indicates the size of that puncture which the fleam can leave behind. It is more safe than the lancet, which, though of a more innocent aspect, has inflicted wounds of awful dimensions. For the last reason, the employment of the lancet by veterinary surgeons is not to be commended.

Above the cutting-point of the fleam, and upon the opposite side of the blade, is seen what is intended to represent a bulging piece of metal. That indicates the place which the operator occasionally strikes with the side of his hand; its intention is to afford a blunt surface for delivery of the blow.

A BLOOD-STICK, WHICH IS LOADED AT THE LARGER END.

It is advantageous to possess a fleam of the above form, because, under rare circumstances, the possibility for which it provides may be encountered; but, for general use, a blood-stick is more instantaneous, and is more certain in its result, wherefore it is to be preferred to the human hand, as giving the smarter impetus to the blade.

A blood-stick is merely a hard piece of wood, six or eight inches long, and turned in a lathe till it has assumed the above form: the larger end is then hollowed; the cavity is loaded with lead. Such a tool, though very diminutive, can be made to deal a heavy blow, and it is quite powerful enough to send the point of the fleam through the skin and thin layer of muscular fibre which externally cover the jugular vein.

However, before any attempt is made to bleed the horse, the animal's eyes should be bandaged. Most animals, from natural timidity, shrink, if they can discern when the blow is about to be delivered, and the point of the fleam is thereby frequently displaced.

BLEEDING.

The sight should be first obscured, then the vessel raised; afterwards, the fleam arranged upon the huge pipe, thus brought into view, when, a sudden blow being dealt with the blood-stick, will cause the current to spurt forth. Should any accident prevent the first attempt from being successful, the operator should not strike twice in the same place. Repeated blows upon the same spot are likely to bruise the part, or to cause a ragged wound, neither of which circumstances is favourable to the healing process. Leave the slight incision

to nature, for it very rarely requires any treatment, and,
choosing a fresh mark, repeat the process with better success.

Blood being obtained by the operator, the groom approaches,
bearing the blood-can. This the man presses against the

PRESSING THE BLOOD-CAN AGAINST THE NECK TO ARREST THE DOWNWARD CURRENT,
AND TO CAUSE THE BLOOD TO FLOW FORTH.

horse's neck, thereby impeding the downward stream within
the vessel, and causing the vital current to gush forth.

Whatever may be the urgency of the business which may
demand your presence elsewhere, never quit at this stage of
the proceeding. However experienced or meritorious the
servant may be, always remain until the operation is concluded. These poor men invariably possess opinions of their
own, that are stronger because of the ignorance upon which

such notions repose. The groom may have seen a gallon, or
even two gallons, extracted, when in his last situation. Such
people delight in strong measures, and will sneer at the one
or two quarts you may desire should be withdrawn.

When the determined quantity has been extracted, remove
the pressure below the orifice, and the outward stream will
cease: then proceed to pin up. Having rendered the point
of a pin somewhat angular, by cutting off the tip, the wire will
pierce the integument the more readily. Drive it through each
side of the wound, and, being in this situation, twist, after the
fashion of a figure of ∞, some tow or thread,
or a hair pulled from the horse's tail, round
its either extremity. Subsequently, remove
so much of the pin as may protrude, and the
orifice will be closed by what surgeons de-
nominate a twisted suture.

TWISTED SUTURE.

When performing this, a few precautions
are imperative. In the first instance, the surfaces should not be brought immediately together. The wound should be left open until the lips become
sticky, as, when in that condition, they unite the more readily.
Next, when closing the orifice, all hairs should be removed,
which is sometimes difficult, should the integument have been
torn asunder with a blunt fleam. The skin then is twisted
and forced from its integrity, but, if a sharp or proper instrument has been used, the presence of hair is never annoying;
indeed, it seldom requires attention.

The sides of the incision should be adjusted with all nicety,
because, subsequently to bleeding, healing, by the first intention, or by the speediest natural process, is desirable. Hairs,
when present, prevent that union from being perfected: they
irritate the part, and act as minute setons, which provoke
suppuration. The advent of the last action is always to be
feared after a vein has been opened. Pus gravitates into the

vessel, and the blood becomes vitiated, when the results are always to be lamented, as they sometimes prove fatal.

After bleeding, care should be taken, until after the removal of the pin, to place the patient in a position so as to prevent the possibility of his rubbing his injured neck, but, at the same time, he must be fed on a digestible and nutritious diet until the system has regained its former vigour.

CHAPTER III.

SHOEING—ITS ORIGIN, USES, AND VARIETIES.

SHOEING a horse is understood to signify, fastening a piece of iron to the horn which envelops the foot of the animal. Such an operation, at first glance, appears to be so simple an affair as to admit of few remarks ; but there is no subject associated with veterinary science on which more research has been expended ; about which more bitter discussion has been indulged ; or with regard to which proprietors and practitioners are more at variance. Certainly no matter can possibly be more intimately connected with the sufferings and the comforts of the equine race.

EARLY ARABIAN SHOE.

The custom of fixing iron to the hoof of the quadruped would seem, at the present moment, to be all but universal. This habit was, probably, derived from the East. In portions of the Desert of Arabia, a primitive-looking shoe is still employed, which, like most things in that region, has possibly remained unaltered during the passage of centuries. Such articles retain the impress of a bygone era, being merely pieces of sheet iron stamped, not forged, according to a particular pattern. The reader may be puzzled to form an accurate notion of such things; therefore, illustrations representing present and ancient shoes are appended. Both partake of the same general characteristics; but, among a people so widely scattered as "the children of the Desert," doubtless numerous variations, as regards particulars, might be selected.

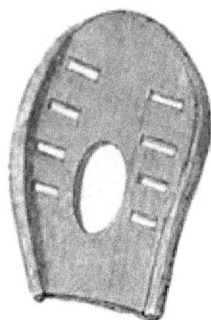

ARABIAN SHOE OF A MODERN DATE.

These shoes look like things produced during the childhood of civilization; but, to assure the reader that at one period horse-shoes resembling the above were almost universal, below are subjoined sketches of those adopted, even at a recent date, by the Moorish, the Persian, and the Portuguese nations. These people are widely distributed; but they are all characterized by the tenacity with which each has clung to the habits of its ancestors. The shape pervading the examples brought forward is too eccentric; the generic likeness is too remarkable, and the peculiarities of feature impressed on each are too conspicuous to permit of their united evidence being pushed on one side with any commonplace reference to an accidental resemblance.

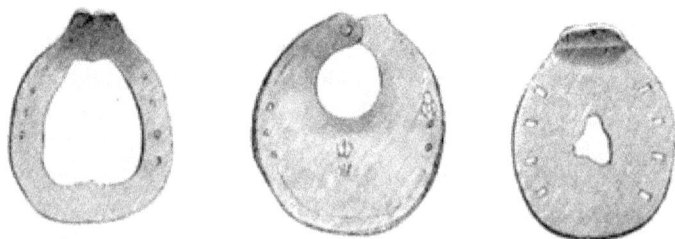

A MOORISH, A PERSIAN, AND A PORTUGUESE SHOE, COPIED FROM GOODWIN'S SYSTEM OF SHOEING HORSES.

Succeeding the above engravings, is appended an authentic

sketch of the old English horse-shoe which was in common
use at the commencement of the last century. When compared with the plate of the Arab, which, doubtless, was the
original, it assuredly exhibits signs of intention. The calkin,
intended to prevent slipping, we here see, as likewise in the
foregoing examples, is by no means a modern invention. The

OLD ENGLISH SHOE. COPIED FROM CLARK'S WORK ON SHOEING.

position of the nail-holes has been materially altered: they
have been moved from the centre, and have been made to
range around the outer margin and made to pierce the solid
horn of the toe, which previously was scrupulously spared.
The fastenings, likewise, have increased in number, having
grown from eight to fourteen. The central opening has been
enlarged; but the thickness of the iron, and the general figure,
however, demonstrate the source whence the original was
derived.

Thin plates of iron were once nailed as shoes to the hoofs
throughout Great Britain. The breadth was not, perhaps,
considered a decided disadvantage, when roads were few, and
much marshy soil had to be crossed in a day's journey. But
if this peculiar form enabled a steed to walk more securely on
a soft surface, the suction, inseparable from such land, must

also have exposed the animal to the frequent loss of the appendage. When regarding these unavoidable results, we can perceive the reasons which have dictated all the subsequent alterations. The central opening had been enlarged, in the expectation of thereby counteracting the sucking effects attending the movements over a marshy country, while the nails had been increased in number in the expectation of thus gaining additional security. The fastenings had, likewise, been ranged round the rim, so that these might be driven directly through the hardest part of, and have longer hold upon the most resistant portion of, the horn.

Such plates were at one time, no doubt, in general use throughout Great Britain ; and, illustrating whence they were derived, there may be adduced a well-known fact. The race-horse is of almost pure Eastern blood. The trainer's stable is a very conservative locality, into which changes slowly enter, and where names are retained long after their applicability has ceased. A thoroughbred is spoken of, to this day, as running in "plates," although the contest is decided in shoes resembling those worn by other animals, only of lighter make, and of the highest possible finish.

The arrangement of the nails near to the outer edge, and the fixing of them into the hard outer wall of the crust, are methods still followed, though experience has demonstrated that such numerous bodies, driven almost perpendicularly into a thin and brittle substance, were better calculated to break the hoof, than likely to hold on that which it was their single office to retain. The modern smith, moreover, does not, generally, puncture the toe of the foot ; but the situations of the nail-holes, and the direction of the nails within such a part, must have been originally regarded as a vast improvement upon the prevailing customs.

That which was formerly an innovation is, however, now the custom. No other mode of driving the nails is, at present,

in general practice, though the modern veterinary surgeon
recognizes all the evils which attend the habit; yet these
evils he contentedly classes as
diseases, instead of seeing in
them the natural consequences of
a faulty system.

In the sandy Desert of Arabia,
where a flat and perfectly dry
country rendered suction impossible, any degree of tension,
however feeble, might serve to
keep the horse's shoe in its

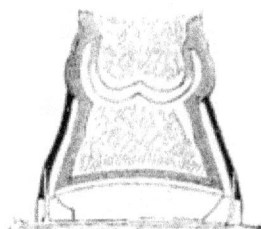

THE PRESENT METHOD OF FASTENING
THE ENGLISH HORSE-SHOE ON TO
THE HORSE'S FOOT.

situation. On such a soil, eight lateral fastenings—each no
stronger than a stout wire,—might afford all needful security.
The size of the holes assure us of the bulk of the nail-heads,
the projection of which, probably, served to give security to
the tread, as well as to retain the metal: being inserted at
one end, and driven with the hammer to the other extremity
of the opening, they might be an ample provision for such a
purpose, when the desert permitted no vast amount of wear,
and the nature of the animal assured lightness of motion.

The English reader may feel disposed to sneer at the Asiatic
manner of fastening the shoe upon the horse's hoof; but he
will do well to inquire "whether the modern method of attaining the same object is altogether free from objection." To
enable him to do this, it is necessary that the composition of
the outer wall of the equine hoof should be explained.

The *wall* of the foot is so much horn as can be seen when
the hoof rests upon the ground, and when it is viewed
either immediately from the front, or directly from the sides.
This wall is supplied from two sources. The coronet, or the
prominence to be seen immediately above the hoof, secretes
the outer layer of horn, which is the darkest, is very much
the hardest, and is the most brittle of all the constituents

of the hoof. The laminæ, or the highly-sensitive covering
of the internal foot, secrete the inward layer of horn, which
is soft, tough, and devoid of colour.

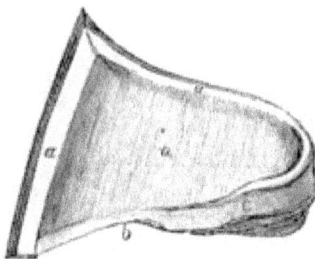

DIAGRAM, ILLUSTRATIVE OF THE DIF-
FERENT KINDS OF HORN COMPOSING
THE HORSE'S HOOF.

a a. The wall. The outer dark portion
is called the crust of the wall, and the
light-coloured, soft, inner horn is
thrown into the laminæ, or thin leaves,
whereby it gains extent of attachment
to its secreting membrane.
b. The light-coloured and yielding horn
of the sole.
c. The tinted, but elastic, horn of the
frog.

These two opposite and
distinct secretions are, by
nature, joined together,
forming one body. Now,
the intimate union of opposite properties endues the
substance, thus compound-
ed, with the characteristics
of both. The hard out-
ward horn was needed to
protect the foot against
those stones and rocks over
which the animal was in-
tended to journey. The
internal, white horn, being
fastened upon this sub-

stance, acted as a corrective to its harsh nature, preventing it from breaking, from splitting, and from chipping,
which it else must have done under the weight it was destined
to sustain, and when fulfilling the purposes to which the
horse's foot was designed to be subjected.

Pathology has indirectly recognized the intention of this
junction, by acknowledging that condition to be a state of

FALSE QUARTER, OR A DEFICIENCY OF
THE OUTER WALL.

THE ONLY POSSIBLE RELIEF FOR
FALSE QUARTER.

disease, wherein the two kinds of horn are separated. Such a
division is known as a seedy toe and as false quarter; and

the foot is recognized as weakened, when such a want of
union is discovered. The outer, dark-coloured horn, becomes

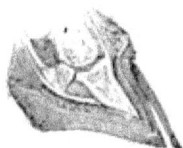

SECTION OF A HORSE'S FOOT AFFECTED WITH SEEDY TOE.   A FOOT WITH SANDCRACK.

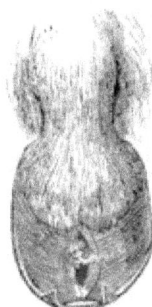

more brittle; the white, internal horn, grows more soft, for
the want of that junction, by means of which each communicated its attributes to the other. So, also, when the two
descriptions of horn, although united, cease to influence one
another, pathology acknowledges this condition as a morbid
alteration, and, when the wall divides from the coronet to the
ground surface, the disease sandcrack is manifested.

The untutored Arab, however, takes advantage of the united
properties of the horn. In warm countries the horse's hoof
grows strong and thick. The Asiatic allows the wall to descend half an inch below the
sole, and, right through the
entire of this portion of projecting hoof he drives the nails
which secure the shoe. Proceeding thus, he does not injure
the foot by the insertion of
foreign bodies through its more
brittle substance, while he se-

THE MODE OF FASTENING THE ARABIAN
SHOE TO THE HOOF OF THE HORSE.

cures the united resistance and tough qualities of the complex
covering of the foot.

The English smith, on the contrary, by ranging the holes for the fastenings round the edge of the shoe, drives the nails only into the harder kind of horn, and transfixes the crust for a considerable distance. The English shoeing nail is meant to pierce only the black or outward substance of the wall. This may, seemingly, afford the better hold; but it also offers the more dangerous dependence. There is, likewise, the peril to be braved of pricking the sensitive foot, should the nail turn a little to one side—an accident which not unfrequently happens. There is, moreover, another danger, namely, that which the forge calls driving a nail "too fine," that is, forcing it near the white horn, rather than sending it directly through the centre of the narrow dark crust.

FRACTURED CONDITION OF THE HORN, CONSEQUENT UPON DRIVING NAILS THROUGH THE BRITTLE OUTER CRUST OF THE WALL.

There remains to be enumerated a third peril. Horses with thin walls present difficulties to the shoeing smith. He is afraid of either pricking the foot, or driving the nail "too fine": should the last accident ensue, the nail will, upon the animal being worked, bulge inward, and provoke acute lameness, often causing pus to be generated. To avoid these evils, he points his nails outward, and, by so doing, not unseldom induces the harsh outer crust to crack; to split up, or to chip off. To such an extent does this sometimes happen, that the smith is occasionally puzzled to find the place where a nail will hold.

Every horseman knows how vexatious it is for a horse to throw a shoe; but the inconvenience to the rider is as nothing when compared with the evils likely to result from such accident when a shoe partially parts company from the hoof. Some portion of the hoof first yields; this throws greater stress upon the remaining fastenings. The shoe becomes loose, and the nails, perhaps, with the exception of one, loose their hold, and the shoe, when raised from the ground, hangs pendulous from the foot, and, at every step, swings from side to side, when the nails remaining in the shoe are prone to impale the plantar surface of the horse's foot, and, in so doing, to puncture the sensitive sole, and sometimes even the coffin-bone, when a very dangerous wound is inflicted.

Against the Arabian method of driving the nails, it may be advanced, that, if the equine hoof is permitted to grow, the elongation of the horn at the toe, and its non-removal by the knife, would occasion this portion of the foot to protrude, and ultimately curl upward like a Turkish slipper, such being the result of long-continued neglect, as is exemplified in the feet of too many donkeys.

THE SHOE PARTIALLY BREAKS FROM THE INSECURE FASTENINGS, AND ONE OF THE NAILS, STICKING UP, PIERCES THE SOLE OF THE FOOT.

It is not proposed to subject the horse's foot to anything like the usage to which the hoof of

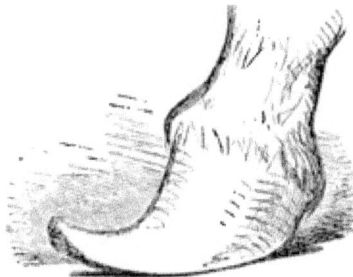

THE NEGLECTED AND LONG UNSHOD HOOF OF AN ASS.

the ass is habitually exposed. All the writer contemplates is, moderating the smith's employment of the drawing-knife, and

6 *

of the rasp, enforcing some caution in the application of the red-hot iron, when burning a seat for the shoe. Why need the wall be always cut away till it is level with the horny sole? Why bring this last portion of the pedal covering, which is naturally soft and yielding, on a line with that part of the crust which is imbued with a power of resistance? Nay, the harder wall is protected by the shoe on which it rests, while the softer sole is brought near to the ground, being left exposed to an injury, which the lesion known as bruise of the sole, proves not unfrequently to happen.

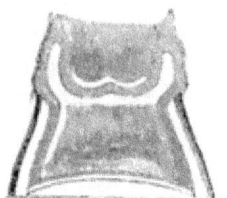

ENGLISH MODE OF PARING THE HORSE'S HOOF.

The sole, being exposed thus close to the earth, is the fruitful source of several "accidents." The soft horn of this region being brought so low, is rendered constantly wet: the consequence is a harshness of texture, perfectly opposed to the evident intent of nature. This harshness is one of the most common sources of corns. The edge of the sole rests upon the web of the shoe: the descent of the coffin-bone, being unable to play upon a yielding sole, squeezes the flesh between the inferior surface of the bone and the upper surface of the shoe. This is acknowledged as the principal source of corns. Stones and other rubbish often become impacted between the horny sole and the shoe. In this situation the foreign substances are retained so firmly, and provoke such acute lameness, that it is common for all stable-men to keep by them, as well as it is general for most horsemen to carry, a curved tool, denominated "a picker." Such annoyances, with many others, might be easily avoided, could the English smith only be prevailed upon not to pare the sole so thin that blood bedews its surface, and then to make the level of the diminished part the point whereto the crust is to be lowered.

Another probable consequence, attending the customary cutting away of the sole, has not been sufficiently considered. The shape of this part, its yielding character, and its position immediately under the coffin-bone, all should be accepted as proofs that it is of service in supporting the weight of the body. It proves nothing to assert that, if the sole is removed, the pedal bone will not fall down. The burden may repose upon the numerous laminae, and upon the bulging rim of the coronet, as well as drag upon the lateral cartilages. Here is sufficient material to uphold even a greater load; but can such a force be arbitrarily imposed by human authority without provoking Nature's resentment? The parts, here named, are the very regions which are the common seats of foot disease. Ossified cartilages,—irregular secretion of coronary horn and laminitis, in the acute or in the chronic form,—are very common to stables; so, also, is navicular disease, which the trimming of the frog is also likely to induce.

Navicular disease, of all other foot maladies, is the one most to be dreaded, and is, in many cases, produced by bad shoeing; consequently, it will be advisable, whilst explaining "how to shoe a horse," to draw attention to the causes in shoeing which have a tendency to produce it. Before doing so, it will be necessary to detail the way in which the horse brings his feet to the ground, i.e., how locomotion is effected; and for this reason, for years previously to 1858, persons thought, and perhaps even now many consider, that, during progression, the horse brings his toe in contact with the ground before his heels: this was the teaching at the Veterinary College during 1857, to which, in 1858, the reviser objected in the following letter :—

"Physiological Reflections on the Position assumed by the Foot of the Horse in the varied Movements of the Limb. By J. I. Lupton, Student of Veterinary Medicine, R.V.C., 1858.

"1. The foot of a living horse, in a state of rest, remains

firmly on the ground —that is, the toe and the heel are on the ground at one and the same time ; but, if during this position the extensor muscles were to contract, then the toe would be raised from the ground ; and if, on the other hand, the flexor muscles were to contract, then the heel would be raised from the ground. Now, during progression, the first movement which takes place is the contraction of the flexor muscles, by which (together with the muscles of the arm) the foot is raised, the toe being the last part of that organ raised from the ground. The foot is now in a position to be sent forward, which is brought about by the contraction of the extensor muscles ; the foot is then thrown out as far as the flexor muscles will admit, and, when at the greatest allowable point of tension, the heel is brought in opposition with the ground. The flexors now in their turn contract, the heel is first raised from the ground, and lastly the toe, which brings me back to the point I started from.

"2. Viewing the leg of a horse as a piece of mechanism (allowing that leg to be even in a state of anchylosis), and comparing it to the spoke of a wheel, during the revolutions of which the posterior part of the inferior extremity, or, in other words, that part which is attached to the tire, comes in contact with the ground first ; but if, in the place of the spoke, the above-mentioned leg of the horse were there placed, the heel in that case would come in contact with the ground first, and the toe last.

"3. As to the anatomy of the foot. The foot is composed of the os pedis, os naviculare, and a small portion superiorly of the os coronæ. Between the aloe of the os pedis we have the frog, and the fibrous frog ; in fact, a beautiful elastic cushion ; and postero—laterally the lateral cartilages, readily yielding on the application of pressure. Seeing this arrangement, I naturally seek to find the cause of its existence ; and I suggest that it is there in order, by coming in contact with

the ground first, to break the concussive effect, likely, if being
hard and unyielding as the formation at the toe, to be productive of much cost to the animal frame.

"4. The progress of action is from the heel to the toe. For
example, man, during progression, puts his heel to the ground
first ; the ox also places his heels similarly on the ground first,
and dogs bring their heels in contact with the ground first :
does it not, then, seem undeniable, when reasoning by
analogy, that the horse similarly brings his heels to the
ground first ?

"During progression the body moves forward, during which
movement the toe, as evident to every observer, leaves the
ground last, that is, when the flexors are contracting. If such
be the case, then, for the toe to come in contact with the
ground first, as some affirm, and the heel last, is a retrograde
and impossible movement.

"Three principal impressions are made on the foot during
progression ; namely :—

"1. On the heel, when great expansion and yielding takes
place, owing to the pressure on the frog, which is forced upwards, causing the ultimate expansion of the walls of the
hoof, &c.

"2. On the middle part of the foot, when the bones bear the
weight of the body, the flexors and extensors being, for the
instant, in a state of quietude, i.e., neither of them are extending or contracting.

"3. On the toe, when the animal gives a push, by which an
impetus is given to send the body forward.

"The foot comes on the ground nearly flat, I admit, but
the heel is for an instant on the ground before the toe.

"I assert, in conclusion, that the progress of action is from
the heel to the toe, and not from the toe to the heel."

My position of the heel progression was warmly contested
on the one side, and as strongly advocated on the other, by

several veterinarians, principally by Mr. Gamgee, senior, on my side, who, about a year previously to this period, in a conversation with me, evidently considered the old doctrine of toe progression to be correct; but he wrote a year afterwards:—"Now, my conviction is that, in the solid-hoofed quadrupeds, the general law is obeyed which guides the movements of the limbs in all animals, and that in the horse the contact with the ground, not only with the fore, but also with the hind feet, occurs, first at the posterior, and last at the anterior part of the foot."

Thus it will be seen how deeply rooted had been the impression in the theory of the toe coming in contact with the ground before the heel, that a certain time was required before veterinarians old in the profession accepted the new statement of heel progression as a fact; but even now some, perhaps, are to be found who may raise objections to it; therefore, being deeply convinced that the question is one of great importance, both in a physiological and pathological sense, both as to shoeing horses as a preventive of disease, and in treating feet when diseased, I cannot occupy the space allotted better than by answering objections advanced against my paper (above quoted) by some veterinarians, by A., who commenced his letter by stating that he did "not attach any deep importance to the subject." Neither did he wish to attempt "even a concise definition of the progression of an animal," but stated "that the operation of muscular power which effected it, though regular, was, in his opinion, very indefinite as to consecutiveness." A. therefore considers the position assumed by the foot of the horse during progression to be a matter of very small importance. Does not the mechanic, in forming an instrument, take care to have every part well adapted to the other, and does he not well know the exact point at which each touches the ground, or any part of the machinery, as the case may be? He does, and it is the duty of the physiologist to ascertain how the piece of mechanism (the horse) works; where each attachment meets, and the exact point at which the foot meets the ground. Again, although an attempt to give a description of progression is not ventured, and although "muscular power that effects it is regular," yet it is very indefinite as to consecutiveness. It must be plain to any reader that this first sentence above contradicts the second: how can the action of one part on the other be indefinite as to consecutiveness, when the movements are regular—they either are regular or not—*Utrum horum mavis accipe.* They are regular, doubtless. In progression we notice, first flexion, secondly extension, and thirdly approximation; and these two latter movements could not be performed previously to the former being put into action.

The reasons assigned for believing that the toe primarily reaches the ground, are:—"The wear of iron in uncalked shoes, in the majority of cases, is most evident at the toe and outer edge rather than at the heels, which assuredly would not be the case if the heels, even for an instant, came primarily on the ground."

The fact (for such it is) of the shoes wearing away at the toes more than at the heels, would be exactly the condition of shoe I should expect to find after three weeks' wear; because, if the heel reaches the ground first, its implantation on the ground is (to use a word) impressive, not the grinding down we notice at the toe, where the action is attritive "when the animal gives a push, by which the impetus is given to send the body forward," and thus a grinding power is continually occurring during locomotion: this state of shoe is commonly noticed in horses who are good movers, and who markedly bring their heels to the ground before the toe. B. considers that it is impossible, in attempting to prove the truth of this subject, to talk about "viewing the leg as a piece of animal mechanism," when its original movements are destroyed by anchylosis. When comparing, for the sake of argument, the leg to a piece of mechanism, viz. a spoke, I merely intended to draw the mind to the fact that the progression of all bodies proceeded from behind forwards. If the anterior part of the extremity of a spoke came near the ground before the posterior, a retrograde movement must necessarily follow; whereas, by the posterior nearing the ground first, progression must result as a consequence. The foot of the leg which I suggested might be placed within the tier, was to be there located, so that the toe and heel should be parallel with the earth when immediately over it, and then told the fact that the heel in progression would reach the ground before the toe, and, moreover, by allowing the existence of anchylosis, the case was made stronger for my argument, for, by assuming this state of things, all the movements which operate in bringing the heel to the ground were removed, and a dead leg as stiff as a spoke was alone left in its place, and with this allowance, and with the fact before his eyes that the posterior part of the spoke reaches the ground before the anterior, and the heel of the foot before the toe, in the leg in which no movements between joints were permitted, B. is struck forcibly —"That the toes of the legs so placed would come first in contact with the ground during the revolution of so 'singular a wheel.'"

Again the anatomical construction of the foot—"Evidently as beautifully formed to resist concussion as it is possible to conceive any object to be adapted to its purpose."

It is assumed that the adaptation of the frog to receive the first effect of concussion does not by any means solve the point, for "I contend," writes B., "that the horn at the anterior part of the foot is as capable, if not more so than the frog, of resisting as much concussion as it may be called upon to bear, considering that concussion is so instantly and thoroughly distributed through the mechanism of the horse's frame."

In answer to this assertion I would ask any reasonable person whether a hard and almost unyielding substance, like the formation of the hoof at the toe, is better able to withstand the repeated effects of concussive shocks than the construction at the back part of the foot, where an elastic frog exists, bounded by yielding sidewalls (lateral cartilages); in short, an adaptation of parts, which any mechanic would with little consideration pronounce to be a surface calculated to ward off concussion, and to be the best contrivance for preserving from injury the superimposed structures. Moreover, it has been said by a philosopher, that when any doubt exists between two questions in the order of nature (such as the one now in dispute), that side is always to be taken where construction shows greater adaptability to a specific object. As to the concussion being "so instantly distributed through the mechanism of the horse's frame," let me ask, would the concussion be distributed immediately through the frame if the toe was the first to receive it ? I think not, and for the reason that the effect of a hard substance striking the ground would produce a shock, whereas a yielding one would, joint from joint, facilitate the distribution of it through any mechanism ; doubtless, therefore, a general law governs and "guides the movements of limbs in all animals," and this law can, as regards progression, be noticed throughout the entire species of biped and quadruped animals, and it consists in action from heel to toe, and not from toe to heel.

The next argument brought against heel action is the example given of men, during progression, bringing their heels to the ground before their toes. This B. considers is "not fair analogy," as the foot of man includes the os calcis (or true hock-bone of the horse) "as well as the digital extremities." "Though heel and toe is fair walking, yet in rapid progression the heels of the swiftest runners do not touch the ground ever so lightly." This assertion is quite true, but it is

moreover equally true that their *toes proper* do not come to
the ground first, even during very rapid progression; and if a
comparison were made between the horse and man when
swiftly running, my opinion is that the former places principally his heels on the ground, whereas the latter brings the
lower ends of his metatarsal (or canon bones of the horse) first
to the surface over which he moves, or, in other words, if the
horse so progressed, he would always, when racing, be going
on his fetlocks; for evidently man uses these joints when
running; consequently I think my analogy was quite fair,
because man places on the ground that part of the foot most
posteriorly situated, and similarly, and for the same reason,
does the horse use his heel before his toe; and if we continue
our observations to other creatures, we find the dog, the ox,
the cat, and numerous other animals, progress from heel to
toe. Is it not therefore very peculiar that the horse alone, of
all quadrupeds, should be singled out as an animal moving on
his toe before his heel, whilst other creatures progress by
bringing to the ground first that part of the foot that is most
posteriorly situated, the dog and cat those parts corresponding
to the fetlocks of horses, and to those portions of the foot of
man, toes, and the ox going on two heels instead of one, like
the horse; but as regards the anatomical analogy by pro-
gressing similarly as that animal; *i.e.*, on the heel of the distal
phalange or coffin-bone?

When horses are going down hill, also, it is readily noticed
that they place their heels on the ground first: the correct-
ness of this assertion B. disputes. He writes:—"See him also
going down hill, backing in his breechings, and checking the
too rapid descent of his load; even here he plants his toes,
fore and hind, first on the ground."

The incorrectness of this view needs no comment; it is
only necessary for my readers to carefully watch any horse
going down hill with a heavy weight behind him, when at

every step he will discover that the heels are markedly and firmly implanted on the ground before the toes; again, the same thing can be easily noticed in the horse who travels even on a level road with calkings, when a tip-tap noise from calking to toe can be heard, in addition to which, ocular demonstration will give a verdict of a first place to the heel; this part of the subject, therefore, will be left to practical observation. We are next asked to "mark the fast trotter as he skims with airy tread the road beneath, the toes scarcely touching it." This I can readily believe; the toes scarcely do touch it—they do for an instant, and that instant is at the moment they are springing from the ground. As in the fast trotter movement is very rapid, so the action from heel to toe is quick, and requires a very keen eye to detect this first movement; the last, or toe action, to the passing observer, of course, must be seen; if not, no part of the feet could touch the ground. The fact of the toes scarcely touching it, proves that the heels first arrive there, for, if the toe reached the earth first, then the heel would afterwards, and by this action tardiness must follow as a consequence, and the rapidity with which the toes left the ground would not be considered so rapid as scarcely to touch it.

The fleet racehorse, also, even "in his most trying moments," puts his heels to the ground first; for, if it were otherwise, what explanation can be given of the mode in which the flexor and extensor muscles act? If the toe touches the ground first, and leaves it last (which latter action all admit), then one set of these muscles would have to perform duty twice to others' once—a state of things totally at variance with the simple means Nature always adopts in attaining her wise ends.

We are informed that in disease the heel action is noticed, particularly in laminitis. This doubtless is the case; but without going into any description of this affection, I think it can be readily understood that even in this instance the example given, in attempting to prove the toe theory, tells the other way; for what part does the creature thus afflicted bring to the ground? That very part where a protective cushion, surrounded by elastic structures, exists, formations calculated to ward off concussion from the diseased foot; in fact, the horse so affected is careful to keep from the ground that very part which B. considers to be better adapted to resist concussion than the posterior part of the foot. The fact is, when the horse is the subject of inflammation of the feet, he goes on his heels, when at every step he protects from jar the diseased internal structures; whereas, if he went on his toes, shock upon shock would produce great suffering, and intensify the malady. Doubtless the horse with navicular disease seemingly goes on his toes, because he is affected with this malady in the posterior part of his foot; he therefore does all in his power to protect the heel from the effect of concussion, which, when there, produces pain; but, even this animal goes on his heel, noticed by the lameness being nearly always apparent.

To insure good shoeing, it is necessary that the farrier fully understands the mechanism of the parts with which he has to deal, and the actions they facilitate; if he does not, he works in the dark, and often labours to create rather than prevent disease, which, as before stated, is constantly produced by bad shoeing. This fact, for so it may be called, was noticed many years ago by the late Mr. Apperley (Nimrod), who, when residing in France, was led to observe that lame horses were very rare in that country, "especially those lame in their feet." This he ascribed to the slow pace at which they then travelled; but a friend of his (a good mechanic) objected to the opinion by stating, "Depend upon it, the French system of shoeing contributes much to their soundness, so far as the feet are concerned, 'by the superior method of nailing.'"

That the French shoe in nine cases out of ten is superior to the English one, no good mechanic will deny — in that it is made to fit the foot, and is supplied with protection where it is needed, and, moreover, is constructed in a manner so as to allow the several functions of the foot to be performed without that unnecessary interference so constantly noticed in the mechanism of an English shoe. A few years ago, when on a professional visit to a large job-master, and driving a horse shod after the French system, my attention was aroused by the remarks "So you are a believer in the French shoe." I answered, "Yes; are you?" He answered, "I'll tell you this, whenever I have occasion to send horses to Paris, those that have bad feet are sure to go, and as sure to return with good ones." This answer will be sufficient now for my readers, as it was then for me. Again, only a fortnight back, a gentleman, a well-known cross-country rider, being one "quite of the first flight" with hounds, told me that, by adopting a system of shoeing similar to the French one, he succeeded in "keeping his horses going," which they failed to do if shod otherwise. From the above, it will be observed that, for years past, practical horsemen have considered the French system of shoeing preferable to the English. Some may ask the reason why the French shoe is calculated to insure the well-being of horses' feet? Possibly it may be because the French mechanic always fits the shoe to the foot; but, in addition to this, the shoe is so constructed as to prevent any portion of the iron pressing upon the sole, except the outside crust (it is well seated out); it is, moreover, a *stamped* shoe, and presents to the ground a *convex surface*, which mechanical contrivance operates most beneficially in warding off concussion, whereas a concave and flat surface both equally tend to facilitate it. The system also persisted in by many of denuding the foot of its natural protection, by opening the heels, as it is called — i.e., in reality, by cutting away a portion of the hoof-horn,

without which the foot must, and always does, suffer; by
paring down horn until the sensitive parts are nearly exposed,
and by removing the larger portion of the horny frog, is most
strongly to be deprecated. It is done doubtless in ignorance,
and for this reason farriers of the present day ought to go
through a course of instruction in the art of farriery, previously to being allowed to operate upon structures, the anatomy, physiology, and economic uses of which they have never
studied, and consequently never understood; such a step
would act most beneficially to the well-being of horseflesh,
and would produce in a few years an intelligent class of shoeing smiths. *Emollit mores nec sinit esse feros.* But to the
subject. This removal of horn—*i.e.* this persistent (from
time to time) cutting away of the horn—leads to the absorption of bone, or, in other words, Nature, ever trying to cure,
and in this instance, after the sensitive parts of the foot have
been nearly exposed by the knife, in order to protect the
coffin-bone from injury, causes it to recede, which she effects
by causing absorption of it. The foot so operated upon by
the knife and by Nature for a second and third time is similarly
treated, when Nature again keeps pace by seeking her remedy
in absorption of the bony structures, until the original foot-
formation is destroyed, the door is opened, and the thief
(disease) rushes in. With a state of foot as described above,
it is not very difficult to understand how easy it is for navicular disease to arise. The shoeing smith does not know,
neither does many an intelligent horseman perceive, to how
great an extent his horse's foot has been injured, until it is too
late, and sometimes not even then, and in most instances he
never recognizes the cause. The cause of disease is discovered
more over the dissecting-table than in the sick-box. The discovery of this disease proves this assertion to be correct. Mr.
Turner noticed it under the dissecting-knife. Veterinarians
of repute never, in their extensive practices, recognized the

cause of navicular lameness until Turner showed them. After-wards they dissected, and specimens upon specimens may be easily seen at our veterinary colleges.

THE SHOE, WHEN SUFFERED TO REMAIN UPON THE FOOT FOR TOO LONG A PERIOD, IS CARRIED FORWARD BY THE GROWTH OF HORN, AND LEAVES THE HEELS UNPROTECTED.

It is well known to physiologists that the constant removal of any natural growth is calculated to result in one of three effects. It may stimulate production, causing the wilfully-excised material to be secreted in unnatural abundance; or, on the other hand, it may interfere with the powers of growth, and occasion the material to be withheld altogether, or cause the product to be secreted in diminished quantity. In such a foot the superabundant growth of horn is owing principally to the diminution of the pedal structures, which arises in such case from the absorption of bone, caused by too close paring of the horn, when Nature, always true to her principles, pro-duces an inordinate growth of horn, in order to make up for the deficiency existing in the internal structures; thus, in rickets, the bones of the leg frequently curve, but Nature strives by extra deposition to strengthen the parts which threaten to break. In the above it has been my endeavour to prove, as regards foot-formation, that any deviation from the natural state is calculated to produce disease. For instance, by removing too much of the heel horn, we cause more stress to be imposed on the tendon of the foot, and consequently greater pressure on the navicular bone, and, by this, produce ultimately ulceration of it, or, in other words, "navicular

7

disease." In making a mechanical instrument, it is necessary
for the mechanic to understand how all parts are adapted one
to another; how irregularity or deficiency in it causes the
whole to fail in operation. So with the horse's foot: he must
recognize, first, how this foot descends upon the ground
during locomotion, and, secondly, how to shoe it so as to
insure its natural action, and protect it from injury when shod.
If the drawing-knife be used sparingly, and with judgment,
benefit may arise; but if it be allowed to cut away the heels,
an oblique soon becomes converted into a right angle, when
the natural bearing surface of the foot is altered, when it can
be easily understood how soon the foot-formation will suffer,
by the production of disease.

Coleman, in carrying out his favourite theory of frog-pressure, instructed shoeing smiths to adapt shoes to horses' feet
in such a manner as to necessitate the frogs being brought
prominently on the ground; and what was the consequence?
Why, the production, in many cases so treated, of navicular
lameness.

No doubt, in a state of nature such pressure gave strength

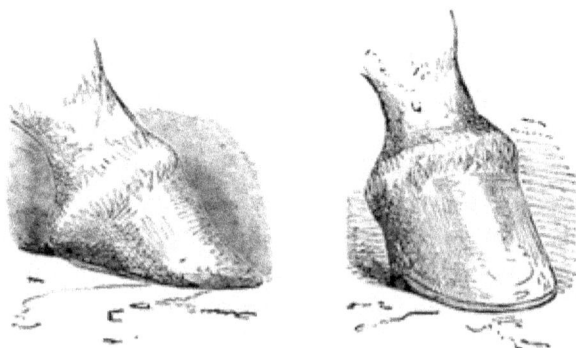

SPECIMENS OF A LOW AND OF A HIGH HEEL.

to the frog, and general health to the pedal structures; but
Nature did not form these parts to be brought in contact with

hard roads, such as we meet with throughout the country, but for soft pastures and plains; in short, Nature arranged a beautifully-made and elastic foot-organization, and formed pathways elastic with herbage for it to tread on; i.e., an elastic cushion, like the frog, was made to meet, when in progression, with an elastic cushion, like the turf. Thus Nature for the wild horse doubly provided for his safety, and removed, by these wise measures, the possibility of injury occurring to the feet; but in an artificial state the case is altogether reversed; the soft pasture is dug up to make room for hard, unyielding stones, noticed in our Metropolitan roads, and around our horses' feet a hard rim of unyielding iron is nailed; and curiously, perhaps, to many, soon after this mechanical application to the feet, the horse goes lame—not immediately, always; not, perhaps, until after repeated visits have been paid to the shoeing forge. It therefore behoves us to consider how to obviate the evils accruing from hard roads and shoeing. The former cannot be altered; without these hard, unyielding roads, our carriages could not be drawn with ease, neither could traffic proceed. Such being the case, all

A SHOE WITH A CLIP AT THE TOE.   THE INCISION, WHICH PREPARES THE FOOT TO RECEIVE THE CLIP.

our attention must be directed to the best mode of shoeing our horses.

7 *

Thick, stubborn hoofs are too common to need comment,
as they are usually produced by lateral nailing; consequently,
by carrying the fastenings to the toe, where the greater thickness of horn affords good nail-hold, we can prevent the occurrence of the ill effects calculated to result from lateral nailing.
The question may be asked, Does the smith respect the integrity of the foot, and particularly the toe horn, when, to
make a cavity for the reception of the clip on the shoe, he cuts
away the horn at the most forward part of the toe? No, he
employs his drawing-knife to cut away horn. Nor is this all:
he sometimes turns up the shoe afterwards to form a calkin,
and actually throws the bearing of the hoof upon that portion
of the foot which he has just denuded of its natural protection.

Calkins, in nine cases out of ten, are needless, and, as they
are at present made, are abominations. The shoe, in the first
instance, is forged too long for the foot, when the extra length
of iron is bent downwards, so as to form a calkin. Below
is a sketch, made from memory, of a very and injuriously
high calkin. It was encountered in the country, soon after
the breaking up of a severe frost; and, probably, it was in-

A HIGH CALKIN.

tended to counteract the wear of
metal which invariably accompanies a frozen condition of the
highways. It would, however,
with a change of weather, fail in
its intent, for the principal wear
is then endured by the toe of the
shoe, and the heel comparatively
escapes friction. All such things
operate according to their height.
They fling the entire bearing for-
ward, where, without any such aid, it must strongly press.
Although contraction of the tendons is mostly confined to
cart-horses (and this constrained position of the foot must

favour such an affection), nevertheless, the smith may receive it as an unjust accusation when he is told that high calkins are to blame for the spread of such a state of disease.

Enough, perhaps, has been said about the evils attendant on the present system of shoeing.

It is argued, and rightly so, that hind shoes require calkings, in order to prevent horses from sliding on our pitched roads, which are met with in all our large towns. Such is the case, but, at the same time, the requirement can be supplied, *i.e.*, a calkin can be obtained, without interfering with the level of the upper surface, on which the foot rests. The shoe can be forged of one thickness from heel to toe, only a portion of metal being removed under each quarter, leaving the calkin to maintain the evenness of the bearing. It may possibly be urged that in thus forming the calkin the quarters become weakened. Nature has, however, set the example by weakening the horn at the quarters; nevertheless, by so doing, she has not destroyed the strength of the hoof. The quarters of an old shoe, when removed, after six weeks' hard wear, invariably are not sensibly diminished in substance, showing that the lessened amount of horn communicates small friction to the metal; besides, the toe is supported upon massive iron, while the heels are upheld by blocks of the same metal. A law of mechanics instructs us that, if the extremities of any powerful substance are adequately sustained, the body which bridges over the space may be without support. The heels being raised to an equal height with the toe, the metal left at the quarters, as it is removed from attrition, is imagined to be fully equal to the necessities of its position.

Most existing shoes are fullered, or have a hollow space, narrow, but long, near to the outer margin. Into this empty void or groove the heads of the nails are received; but, as the substance in front is ground down by wear, of course the duration of the shoe must be shortened in proportion to the depth of the fullering. That the reader may fully comprehend the signification of a fullered shoe, appended is a copy, made from Mr. Godwin's excellent work on Shoeing.

A SHOE, WITH THE NAILS COUNTER-SUNK.
(Ground surface.)

A FULLERED SHOE.
(Ground surface.)

By inspecting the above illustration, which represents the ground surface, the reader will perceive an indented void near to the outer margin. Behind this indentation, or fullered cavity, the iron gradually slopes away, so that the substance which is exposed to wear, and on which the horse must travel, consists of the narrow strip that extends round the outward edge of the shoe.

The proposed shoe contemplates iron of an equal thickness at every point which is usually exposed to wear. The nails are driven into holes made to fit close around the heads of those fastenings; so that, the shoe being fixed, no loss of substance is to be detected, for the nail-heads fill the spaces which were counter-sunk for their admission.

The nails pierce the toe of the proposed new shoe. This part is selected because this portion of hoof is covered with the thicker horn, therefore is indicated as the region where all stress should bear. Among smiths there is a strong objection to driving nails in the centre of the wall. But is not a more violent outrage inflicted by actually removing a portion of its substance, so as to make an abiding-place for a clip, than by piercing obliquely the strongest part of the hoof, subsequent to the toe having grown below the true foot?

The thickness of wall there offers several advantages, when considering the retention of nails. The solidity of the secretion is a proof that this portion of the hoof is not endowed with motion; consequently, when fastening a piece of iron to it, we are not fearful of interfering with the exercise of a

DIAGRAMS, SHOWING THE DIFFERENCE BETWEEN FULLERING AND COUNTER-SINKING, FOR RECEIVING THE NAIL-HEAD OF A HORSE'S SHOE.

Fullering, or a free indentation round the shoe. This space is indicated by the dark portion of the diagram, and within which the heads of the nails repose.

Counter-sinking, or only removing so much metal as may be filled up by the heads of the nails which are to retain the shoe upon the horse's foot.

healthful function. Such would be the case if the nails were to fix the quarters where the joint thinness, moisture, and elasticity of the horn afford the best evidence Nature meant should reside expansion and contraction.

When the contents of the foot are compressed by the superimposed weight of the animal, or when the hoof is resting upon the ground, the quarters yield to the downward pressure, and they accordingly expand. When the burthen is removed by the hoof being raised, the quarters again fly back to their original situations. The sides, therefore, being in constant motion, are entirely unsuited for the purposes to which the smith compels them. No wonder the clenches are loosened, or the shoes come off, when the nails are driven into parts hardly ever at rest: this action is important to the circulation, for the contraction still allows the arterial blood free ingress, while the expansion permits the full return of the venous current.

Therefore, because the thickness of horn denies the possibility of movement; because the amount of inorganic secretion likewise presents a reasonable hope of not injuring other

and more delicate structures; and because the toe affords
those numerous properties, which for the retention of the
fastenings are rendered imperative, the nails, in opposition to
the usage of ages, and the experience of thousands, are fixed
within the anterior of the hoof, seven or five being there
employed to fix the shoe.

Every one knows how much the speed of the racer is dependent upon that elasticity of the foot with which the quarters
are endowed. Who can have failed to notice the play of the
fetlock by which the "blood action" is characterized? Now,
Nature never forms one part an exception to the whole. She
delights in harmony; consequently, the spring which resides
in the fetlock is positive evidence of the elasticity which
belongs to the unfettered foot. But the bounding property
which the frog, sole, and quarters would naturally provide,
the trainer counteracts, in order to impose a dangerous article,
which is not a horse-shoe, nor even a respectable substitute
for one.

It is so formed, however, as to exercise the worst functions
of the regular shoe. It is a fetter upon the foot, and firmly
impales the quarters, thereby seriously crippling the animal,
and impeding the natural power. If any part of a thoroughbred's foot required metallic protection, it could only be the
toe, for this part, alone, is hard and unyielding. The other
portions of the hoof touch the ground first, when aiding to
prevent concussion; but these are never used with that
amount of energy which necessitates anything approaching
artificial defence. Now, the plate and its nails check expansion; these also oppose that force of rebound residing in the
hoof and in its various structures. The best horse must feel
the bondage most. The spring, or rebound, is, to it, of
most value; but that function is destroyed. Many a fine
animal has, doubtless, been condemned for having " no go
in him," which, could it have exerted all its natural power,

would have been declared winner of every race for which it was ever entered.

The late William Percivall, the respected author of Hippo-pathology, many years ago stated that he had long ridden a young horse about town with no greater protection to its fore

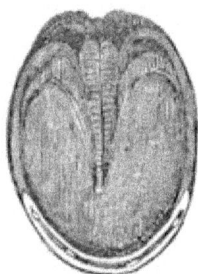

A TIP, OR HALF-SHOE, NAILED ONLY TO THE TOE, AND LEAVING BOTH THE QUARTERS FREE.

feet than tips could afford. He showed the hoofs of the animal, and more open or better examples of the healthy horse's feet need not be desired. Why could not tips be employed by racers, instead of the present ridiculous pretence at a shoe? If any greater protection is imperative, or is thought to be needed, the shoe proposed above would give all security, while it left the pedal structures free to exercise their important uses. There can be no doubt as to the safety of tips; in which, if Mr. Percivall could for years take his quadruped through the streets of London, another animal might, surely, scamper over the well-kept turf of a racecourse, where the heels merely touch the earth impressively, in order to break the first shock when descending on the ground, and then only for an instant.

Were tips more generally employed, this form of shoe would be more highly valued. They are, however, now thought only to be of service when the animal is, "for a season," thrown up; but there can be no reason why the racer—trained, exercised, and worked always on choice turf—should ever be crippled by any more regular form of shoe. Most horsemen,

however, like the warriors of old, place their great dependence
on the accumulation of iron. The nearest approach they ever
make towards a tip, and then only when guarded by a veterinary surgeon's advice, is a three-quarter shoe. The tip is a
protection to be worn only during the run at grass, and to be
discarded so soon as the stable is entered. Is not the racer
always at grass, since the rail or the van generally carries it
over the roads? How often do the feet of the thoroughbred
fail, though there must be further cause than the work they
have undergone? But no one is silly enough to suspect the
shoeing can be at fault!

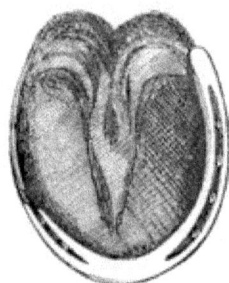

A THREE-QUARTER SHOE, WHICH ONLY LEAVES ONE QUARTER UNFETTERED.

The three-quarter shoe is but an enlarged kind of tip. Most
horsemen appreciate the unilateral nailing which was revived
some years ago by that excellent veterinary surgeon, Mr.
Turner, of Regent-street. They can understand the advantages of leaving one quarter without nails, so long as the
unfettered part be covered by a regular shoe. They comprehend that, by omitting the nails on one side of the hoof,
that side is left free to exercise its natural property of expansion. Therefore, they perceive that the unilateral mode of
shoeing is a partial remedy for contraction.

Though always worked on grass, and ever lightly shod, no
animal is so troubled with mule hoofs as is the racer; yet no

quadruped is so entirely under the inspection of man. The mode of shoeing must be at fault. That cannot be right, the results from which are purely evil. The consequences experienced from the custom of fettering that portion of the foot on which the pleasure of motion and the extent of the rebound both depend, argue strongly in favour of tips, not only as training, but more especially as running shoes. Men with fleshy feet, having no protection from leather, fearlessly tread the racecourse; yet the owners of blood stock seem afraid of trusting their animals to perform an act not equally bold, although Nature sends the horse into the world with ready-made and stout-made shoes. There can be no just reason why the steed which never quits the turf need be hampered even with a unilateral shoe, were the horn only carefully, and not ruthlessly, cut away.

A SEATED AND A UNILATERAL SHOE.

A seated shoe implies a regular shoe which has only so much upper surface left as will admit of the crust resting upon it: the remainder of the web slants away till the posterior or inner margin becomes a comparatively fine edge.

Mr. Bracy Clark once brought forward a jointed shoe, which was intended to admit of expansion, and was offered to the public as a radical cure for all the evils to which the foot of the horse was liable. The joint was placed at the toe, the shoe being forged in two halves, which were united by means of a rivet. The thing was wrong in principle. The toe, which Nature intended should be fixed, was obliged to move, before the heels could expand; then, parts could not yield in dif-

MR. BRACY CLARK'S JOINTED SHOE.

ferent degrees, but all must move at once, according to the motion of the iron. It was soon discovered to be terribly injurious when brought into use. The battering speedily fixed the central rivet, and afterwards wore away the joint, leaving the two halves disunited. A thing which turns out defective,

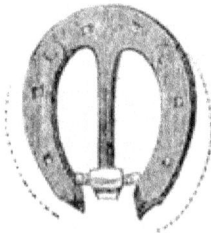

A SCREW SHOE.

both in principle and in practice, merits that neglect into which the jointed shoe has now fallen.

Another mechanical ameliorator was termed the screw shoe. This had two rivets—one on either side of the toe, operating on two movable quarter pieces. The sides, therefore, were capable of all motion, and, being nailed to the quarters, were, by turning the screw, to be forced outward. The screw was situated under the frog, and was retained in its position by a stout bar of iron connected with the toe-piece. Man, however, cannot treat any portion of an organic frame as if it were an inorganic substance. He may tear flesh, but he cannot stretch or strain living tissues according to his pleasure. Moreover, all outward secretions are regulated by the parts which they cover and inclose. Thus, supposing a lad born with a diminutive head, the cranium cannot be enlarged by any degree of force; but, educate the boy, exercise the intellect of the youth, and, with the greater development of the brain, the bones of the head will sensibly expand. So it must be with the heels of the horse's foot. These parts may become rigid and wired in by the fixing power exercised by the nails of the shoe. But remove the nails, allow the hoof that motion which is needful to its health, and its internal structures may recover their lost functions: a gradual restoration to the normal shape may be the consequence of strength regained by the internal organs.

The veterinary mind was, however, slow to recognize so plain a rule. Like all Nature's laws, the truth necessitated not that show of mystery in which the ignorant especially delight. The famous screw shoe is everywhere admitted to have been a decided failure; nevertheless, the pride of poor humanity could not relinquish the hope of compelling life, through the power, to direct mechanical force. Screws and rivets had proved alike hurtful, but there still remained other artifices, which were as yet untried. The frog-pressure shoe was one of these, which ultimately lamed many horses without having benefited a single one. The wedge-heeled shoe is, however, sometimes met with at the present day: it consists of a shoe seated out upon its upper surface, having the heels much thicker than the iron at the toe, the inventor never suspecting, that, when he fixed the quarters of the hoof at a high altitude, and invited the heels to slide down an inclined plane, he was only laying a trap for loosening the clenches, since the quarters and the heels being continuous, one cannot move without the other being displaced.

All men having, theoretically, insisted on the necessity of permitted freedom of motion to the quarters, in order to secure the health of the foot, the next novelty was a proposition to confine those parts by establishing a large clip at either side

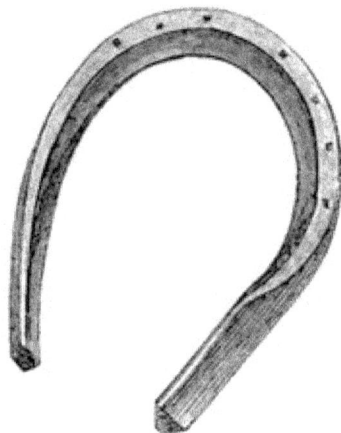

A WEDGE-HEELED SHOE.

of the shoe. The clips were forged, but the thin heels were also retained. The highest portion being at the toe, of course, the foot, obeying the laws of gravity, had an inclination to drag towards the lower level: thus the thin heels had a tendency to draw the hoof away from the clips, one part counteracting the other. Then the clip shoe has a piece of steel inserted at the toe; but, could an everlasting horse-shoe be produced, it would bring but small gain to the proprietor, since the natural growth of the horn necessitates that the metal should be removed, that new nails should be inserted, and that the foot should be pared out every third week: how-

ever, the steel toe and the thin heels were incompatible with
each other, since the thin heels took
the bearing from that part which the
steel presupposes to be alone liable to
attrition.

It would, however, be vain to review all the shoes which have come
before the public. A certain rim of
iron has been pinched up, flattened
out, squeezed in, twisted about, has

A CLIP SHOE.

been lengthened, and has been shortened—subjected to every
species of treatment but the right—and each trivial alteration
has been patented to the public as a final and a wonderful
improvement. After all the many changes, at the present
time a modification of the shoe originally introduced by Clark,
of Edinburgh, is in general use, or, if such an assertion requires any qualification, the hospital shoes, or shoes suited for
particular forms of disease, are the principal exceptions.

The generality of grooms will undertake the relief of those
injuries occasioned during motion, or which are produced by
one leg being hit by the opposite foot.

Of cutting, there are two descriptions. One is spoken of as
"brushing," and this kind occurs near to the pastern joint.
The other is called "speedy-cut," and it takes place immediately below the knee. Both are equally annoying, but the
last is the most dangerous. "Speedy-cut" will destroy the
rider's security in his horse, for, a blow on the seat of injury
may bring the animal suddenly to earth. Both affections are
likely to occasion exostosis, for the repeated injury may so irritate the bone, as shall cause it to enlarge or tumefy : thus the
renewal of the accident produces a result which must increase
the probability of its recurrence.

Almost all weakly, long-legged, and narrow-chested horses
cut. Many young horses strike in going, but they lose

the habit as age matures the strength. Nearly all animals, when exhausted, will "brush," and often very severely. Lately, a ring of india-rubber has been employed as a protection against this annoyance, but it is a mere fantasy, and one not at all calculated to realize any practical expectation. Confirmed disappointment engenders a feeling allied to desperation; but, when nostrums fail, advice should then be sought from more trusty councillors.

Some horses will only cut, or brush, during the latter portion of a long journey, or when thoroughly exhausted. Other quadrupeds are afflicted with a chronic description of weakness, and such animals may cut with the first step. These creatures require less work, or entire rest, with a course of tonics, both in food and medicine.

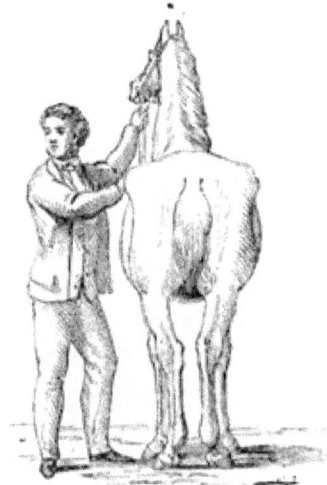

THE KIND OF HORSE NOT LIKELY TO CUT.

However, make and shape certainly have some control over this affection. The horse which exhibits a wide chest, and stands with the feet not too close together, very rarely speedy-cuts. The animal which possesses well-made haunches, with

prominent hips and swelling thighs, that appear full, round,
and fleshy, especially when such a creature places the fetlocks
under the hocks, must be driven very far, and pushed very
hard, before the pace shall become injurious.

Several reputed remedies have been sold for the relief of
this defect. Saddlers keep in stock pieces of leather, or small
flaps with straps appended, which last, being buckled round
the leg, hang pendulous, covering the wound. Such applications, however, rarely are satisfactory. The horse, during the
motion of the feet, repeatedly kicks the leather, and the frequent blows generally remove it from its original situation;
thus, long before the journey has ended, the remedy hangs
over some sound part of the leg, and the sore is bleeding from
renewed injury.

. A better plan is to procure a piece of cloth which matches
the colour of the animal, and to fold this round the leg, ultimately tying it at the top and the bottom. Such a contrivance
cannot be displaced, and is less likely to attract attention than

REMEDIES FOR CUTTING.

the leathern flap recently alluded to. However, it must be
tightly wrapped round the shin, or it will bag, and appear
unsightly, as it is represented in the previous illustration;
still, such a resort affords but a partial protection, cloth being

unable to stay the entire consequences of a blow, nor can it
be regarded as exercising a curative influence.

That which appears better is a leathern boot, of the colour
of the skin, or made of prepared horse-skin, having the hair
on, and laced upon the member. Over the seat of injury a
concave piece of stout leather is let into the covering, and the
hollow, thus formed, which acts as a protection, can also
receive a portion of lint saturated in the lotion, prepared by
adding one grain of chloride of zinc to an ounce of water.
Thus, while the sore is spared a renewal of the cause, curative
treatment is not stayed.

The chloride of zinc lotion is the only remedy which an
ordinary case of cutting would require; but aggravated instances of this annoyance will also be benefited by rest and
a course of restoratives to amend the constitutional debility.
Other matters consist in a warm lodging, an ample bed, nu-

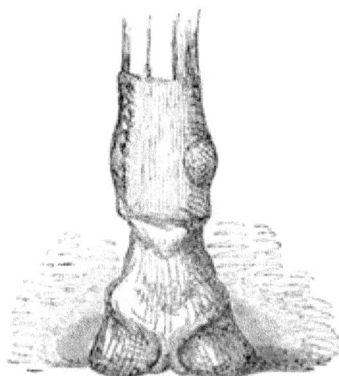

A LEATHERN BOOT.

tritious food, walking exercise, a loose box, and, above all
things, no work. Should the animal be changing its coat,
which is generally a period of weakness, throw it up till Nature's
work is completed; give extra nourishment and one ounce of
liquor arsenicalis, each day, to assist nature. Two months'

run at grass will also be attended with marked benefit, *i.e.*, place the animal for a short period in a state of nature, where succulent food and fresh air can be obtained, and where the system will have time to regain strength, the lack of which, perhaps, has been the cause of cutting.

Cutting is often combined with clicking or forging, for both words signify the same act, implying the noise made by striking the toe or quarter of the hind shoe against the metal nailed to the fore foot. This sound is not generally considered pleasant by those who hear it, because, besides being of a monotonous character, it announces something to be the matter; either that the horse is not exactly in proper working condition, or that the journey has been a trifle too long for the strength of the animal, whilst the repeated blows endanger the retention of a fore shoe.

SHOE, DESIGNED WITH THE INTENTION OF ERADICATING CUTTING, AND OF RENDERING CLICKING AN IMPOSSIBILITY.

The smith, generally, is consulted to cure this defect. He, however, who regards the cause, will perceive that the eradication of the evil more concerns the stable than the forge. The man of the anvil, nevertheless, will put on a novel kind of shoe, which, with all the confidence of ignorance, he shall assert must stay the annoyance. The remedy totally fails, and the horse is led to another forge. The new blacksmith picks up the foot, and, of course, is cunning enough to profit

8 *

by what he there perceives. A different shoe is tried, and pronounced an absolute remedy; still, this disappoints: the quadruped seeks some other shoemaker. The next bit of iron leads to no new result. The clicking and the cutting only get worse during these numerous trials, till the proprietor becomes alarmed, and the horse is thrown up to undergo regular curative treatment.

The rest, thus obtained, often effects that which no change of shoe could accomplish. The smiths, however, are only to be blamed for pretending to perform impossibilities. The best veterinary surgeons in the kingdom, having no better appliances, could have laboured to no better result; the fact being, that the kind of shoe which shall answer in all such cases, does not, and cannot exist. That article has the best chance which is adopted when the owner deems it necessary to lighten the work of his exhausted servant : thus it is a matter of uncertainty which shoe will succeed. The first smith may, or perhaps the last will, prove the very clever tradesman in his employer's estimation.

The next engraving is a type of the shoe commonly employed for the alleviation of this unpleasantness. The number of altered shapes and adapted peculiarities is infinite; but one pervading model is readily detected through all such modifications. There are, however, several shoes claimed as inventions by different smiths, and each is warranted to cure the most aggravated case of cutting or of clicking, on the first application. These shoes frequently fail, while the ordinary shoe often answers admirably, provided the horse be "up" to his work, and not pushed too far, or too hard.

SHOE, MEANT TO PREVENT CLICKING.

The fact being that flesh and blood, if overtasked, will flag, and no mechanical contrivance can anticipate the natural consequences of such exhaustion. Clicking and cutting are not local ailments; therefore, though they may be mitigated, they cannot be eradicated by any local application. They, doubtless, are both produced by the irregular movement of the feet; but the motion of the extremities is regulated by the condition of the body. If the reader is ever on a journey, and the horse he is guiding chances to click, the bearing-rein should be let down,—if the driver ever sits behind harness disgraced by such an instrument of folly. Should that not succeed, accept the warning; pull up at the next tavern, and have the quadruped taken from the shafts, rubbed down, and rested.

After a couple of hours spent by the traveller in the coffee-room, the journey may be resumed, though, of course, a longer stay will rather benefit than injure the steed; yet, in either case, the subsequent pace should be a little slackened, and if, on reaching home, the work is slightly lightened, the noise may never after startle the "ear of propriety."

These remedies should always anticipate the setting in of winter, because wet roads necessitate heavier shoes, by which a severer blow can be inflicted; nevertheless, the majority of horse-owners are extremely careless about the necessities of the seasons. The winters, in this climate, are more generally characterized by their severity, than remarkable for their mildness; yet the frost appears always to take horse proprietors by surprise. Gentlemen, to be sure, during this season, allow their dumb servants to remain within the stable; but quadrupeds which have to work for their own and their master's sustenance; creatures which have to labour long and to labour hard; slaves which toil before the sun has risen, and never cease till darkness has long set in, are never prepared for the season which in England seems a certainty.

A horse-shoe is, however, not a perishable commodity, nor does its store necessitate any sacrifice. Supposing it were forged in the summer, and, because of death or change, it should not suit in the winter, the smith, at such a period, would gladly accept its return. Many forges are comparatively idle during the warmer months, and any amount of winter shoes would be most thankfully manufactured. Then, no one will employ the men; but, scarcely does a severe frost or the snow set in, than people throng into the forge, all clamorous to have their horses' shoes suited to the weather. They crowd the building; they even stop the roadway. The inside is full of men and horses—horses and men cluster deep about the entrance. The smiths have to work fast, and often hang over the fires for three nights and three days without looking on a bed. Beer is abundant, but nature cannot labour continuously on any amount of stimulant, and the men ultimately sink exhausted, to sleep soundly on a heap of old rusty horse-shoes, while many voices are shouting, and many anvils are ringing around them.

Such scenes might be prevented, and the work much better done, would owners lay in a stock of shoes, properly frosted, against the coming winter. The labour executed during the leisure portion of the year, would not be hastily performed by over-taxed workmen; the only extra charge such a provision would necessitate, is the interest on the slight cost of the articles supplied, though very often even such an increase of expense would be avoided, since it is by no means uncommon for the smith's account to remain longer than six months before it is liquidated, while the confusion, loss of time, and those accidents which often occur, would be banished.

Frosting, or roughing, as it is termed, is generally performed in a coarse and careless manner, because of that excessive press of business amid which it is executed. In the first place, the shoe is hurriedly torn from the hoof, without the nails being properly unclenched, or any trouble being taken about the process. Should the proprietor expostulate, he only elicits an uncivil reply, for the journeyman is vexed with boisterous solicitations from a crowd of impatient customers, and irritable from inordinate fatigue. The shoe is then heated, after which, the free extremities are turned downward with the hammer, and the ends are hastily beaten into a rude, sharp edge. In some particular cases, the toe is likewise favoured by having a clip forged, but, occasionally, the toe is turned downward, forming a third and a front calkin.

The rudeness of the above process has long been appreciated by the more reflective portion of the public. To rectify it, various innovations have been proposed. The meditated improvements, however, have all sunk into disuse, because of the attendant expense, or of the necessitated exertion. A common man thinks it no trouble to remain through the night in the blacksmith's forge, waiting for his turn, because other people do the same. But when his turn arrives, perhaps a new set of shoes is spoiled, for the ordinary "roughing" is, generally, of no service after the third day, the sharp calkins being by that time ground blunt.

The huge weight of the animal grinds the edges off the iron, especially upon London stones, so that in three days they are no better than ordinary calkins, and cease to enable the quadruped to progress on ice. The constant removal and renewal of the shoe—the horn each time having to be re-pierced by fresh nails—seriously injures the hoof, so that, frequently, animals are forced to remain idle because there remains no more horn on which to fix a fastening. Those horses which escape such a fate, nevertheless, carry the scars which commemorate the period of frost for months afterwards; for there is no horseman, who has the most trivial experience in such matters, but will bitterly complain of the damage done to the quadruped's feet, when it is forced to work through the winter season.

Some person, many years ago, proposed to use nails with large steel and sharp-pointed heads, during the prevalence of frost. This plan was tried, and signally failed. The constant renewal of the nails was found ruinous to the hoof; for the strongest of the projecting heads was unable to resist the grinding action of a horse's foot longer than twenty-four hours. Then, many of the heads broke off while being driven, and not a few were fixed in a damaged condition, owing to the blows received from the heavy hammer of the smith.

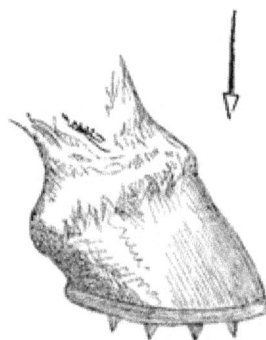

A SHOE, INTENDED TO ENABLE A HORSE TO WORK IN FROST, WHICH IS FIXED ON TO THE FOOT WITH SHARP-HEADED AND PROJECTING NAILS.

Mr. White, however, proposed a plan, concerning the utility of which the reviser is able to bear favourable evidence. Large holes, containing the thread of a female screw, are made through the heels of the winter shoes, and several steel points manufactured with a male screw, adapted to the dimensions of the holes just mentioned. Whenever frost coats the roads with ice, all that is requisite a boy might perform. The hole in the shoe has to be cleared out, and afterwards, with an instrument known as "a spanner," one of the points, before alluded to, is screwed into the opening. When these points are worn down, they are easily renewed ; thus the terrors of the frost are overcome without exposing the horse for hours to the chilly air, or yourself submitting to the incivilities of the forge.

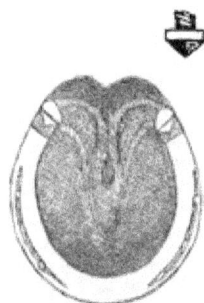

A SHOE, WITH POINTS, WHICH SCREW ON AND OFF, DESIGNED TO FIT A HORSE FOR WORK DURING FROSTY WEATHER.

On the above subject, the following was written in the *Field* during 1861, and explains the measures necessary to be pursued :—

" About this time last season we inserted in the *Field* an account of the plan of frosting horse-shoes, recommended more than fifty years ago by Mr. White, veterinary surgeon, of Exeter. Since then, nearly one thousand sets of the sharp cogs used for this purpose have been sold by the engineer to whom we intrusted the task of making them, and the plan appears to give unqualified satisfaction. At the suggestion of several correspondents who have not seen our former article, we are induced to repeat the notice, with the addition of an engraving representing the tools necessary ; these being, a drill of the required size, which every smith possesses, and with which a hole is drilled in the heel of each shoe, and, if needed, in the toe also. These holes are then converted into

female screws by means of two taps (figs. 1 and 2), one being
slightly smaller than the other, so as to make a perfect female
screw, by using first the smaller one, and then the larger.
Besides these, a spanner (fig. 3) is required to fix on the cog
firmly; and the cogs themselves (fig. 4) should be made by a
competent smith. These may all be obtained of S. Morris,
50, Rathbone-place, Oxford-street, London, the price of the
tools being six shillings, and of the cogs, three shillings per

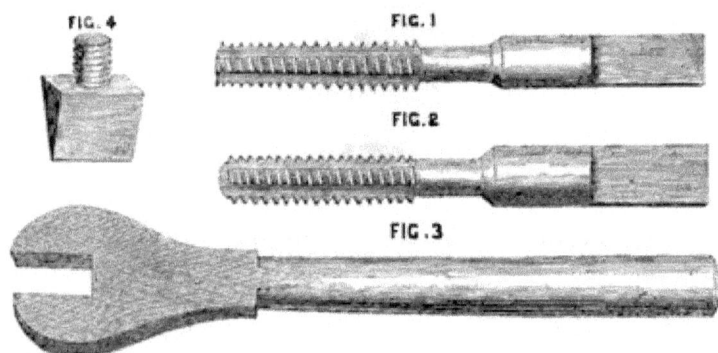

FIG. 4   FIG. 1   FIG. 2   FIG. 3

dozen. With this outlay, any shoeing smith can fit a set of
shoes by drilling the heels (and the toes, if the roads are very
slippery, but for ordinary work the cogs in the heels are quite
sufficient), tapping them with the taps furnished to him, after
which they are nailed on; and the horse so shod can in five
minutes be roughed by his groom, by screwing a cog in each
hole, with the aid of the spanner. It often happens that the
roads become frozen after a horse leaves home; but if the
groom has the spanner and cogs in his pocket, he is independent of the smith, and neither the delay caused by
'roughing,' nor the danger from its omission, is incurred.
A specimen shoe, properly fitted, may be seen at the office of
the *Field*."—*December* 20, 1861.

The plan is excellent, but it requires a little forethought, and a slight expenditure of ready cash. The tools for the tapping, or making the female screw-holes, and for the points, are obtained from Birmingham; the former at a cost of five shillings, the last for one penny, or three half-pence each. Tapping a set of shoes is by the smith charged fourpence; and for so small an outlay the gentleman, just named, escapes the unpleasantness and the annoyance which are inseparable from the old method of "roughing" horses during frosty weather.

The reviser believes he has now touched upon all the *necessary* heads connected with the subject he is at present considering; still, this article cannot be closed without apprising the reader of a practice, not unusual in some forges, but never indulged in by the respectable tradesman. This is,

A FOOT PROPERLY SHOD, AND A FOOT WHICH HAS BEEN CRUELLY RASPED, TO MAKE THE HOOF SUIT A SHOE THAT WAS TOO SMALL FOR IT.

paring and rasping the horse's foot till it be small enough to fit the shoe, rather than kindle a fire and forge a new set which shall suit the feet of the animal. It may to some readers seem like a jest, to write seriously about the horse's shoes being too tight; but it is, indeed, no joke to the quadruped which has to move in such articles. The walk is strange, as though the poor creature were trying to progress, but could obtain no bearing for its tread. The legs are all abroad, and the hoofs no sooner touch the ground than they are snatched up again. The head is carried high, and the countenance denotes suffering. It is months before the horn is restored to its normal condition. The animal must, during this period, remain idle in the stable; and, that the reader may be enabled to recognize the foot, under such circumstances, the last illustration has been introduced.

The horse is so entirely given into the hands of man, and is so submissive to his treatment, that the active supervision of his master is doubly necessary, both for its protection in the matter of shoeing, and in every other detail of stable economy.

CHAPTER IV.

THE TEETH.—THEIR NATURAL GROWTH, AND THE ABUSES TO WHICH
THEY ARE LIABLE.

THE horse was sent on this earth provided with every apparatus necessary to crop, to comminute, and to digest the green verdure of the earth. Man has seized on and domesticated the body, which is exquisitely adapted only for special purposes. He works it while in its infancy, or forces it to labour until the sight is lost, and the limbs are crippled. To fit the creature for his uses, he changes the character of its food. Artificially-prepared oats and hay, with various condiments, are used to stimulate the spirit. No one inquires whether such a diet is the fitting support of the animal. But, when the energy lags, beans, beer, &c. &c., are resorted to as restoratives for exhaustion. The quadruped, thus treated, men have agreed shall be aged by the eighth year; but the reviser has seen very old horses, which had not attained the fifth birthday. Opinion seems to be based upon the circumstance that, by the time recognized as "aged" in the equine species, the indications of the teeth do no more than tempt a guess. The cessation of dental growth, however, does not announce maturity to be consummated; but man appeals to the teeth as corroborative of his judgment, without asking himself whether those parts have been doomed to unnatural wear, and therefore may not have assumed an unnatural aspect.

We have not lately seen a specimen of bishoped teeth. In

Ireland, such sights obtrude themselves at every horse fair. The majority of horses are, in that country, sold cheap, most of the purchasers being clothed in rags. It is a sad feature in the practices of imposition, that it is always violently rampant where there is the least certainty of reward.

To fully explain in what "bishoping" consists, it is necessary to inform the reader, that on the nipping or cutting surfaces of the young horse's front teeth, there mostly are dark indentations, or deep hollows. On the next page is an enlarged engraving of this portion of a tooth, taken from the head of that which was a three-year-old colt. The dark spot, in the middle of the diagram represents the situation of the hollow into which the food naturally falls, rendering the interior of the cavity of a deep colour approaching to blackness. Bishoping supposes the cavity always to be present, invariably to be of one form, and in every instance to sink to the like depth, which suppositions are contrary to fact; but, even were such rules observed by Nature, there are still means by which the cheat may be detected. Immediately around the dark-coloured space is developed a fine line of enamel, which is always white. The rogues can counterfeit the black mark, but they cannot imitate the crystalline white bordering, which surrounds the opening. The presence or absence of this is of more importance, therefore, than the existence of a black indentation. Again, those who tamper with the teeth, cannot change the shape of the surface on which they work. The young tooth is wide from side to side, and narrow from the front to the backward margin. He who ventures where bishoped horses are to be found, should familiarize his eye with the shape of the youthful organ.

In contrast with the natural tooth, the reader is also presented with an exaggerated sketch taken from an organ which had been tampered with, and which was extracted from the head of an animal that had at least attained its twelfth year.

The natural size has been considerably enlarged, it is thereby hoped to render the contrast the more obvious. This last member, it will be remarked, has parted with its juvenile width, or is now characterized by depth and angularity. The central cavity, it will also be observed, bears small resemblance to the natural depression which it is meant to imitate. The colour, moreover, is quite black, and of an even tint throughout, while the presence of the girding line of enamel cannot be detected.

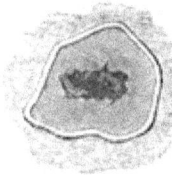

TWELVE YEARS OLD.   THREE YEARS OLD.

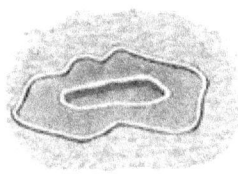

AN ENLARGED VIEW OF THE DIFFERENCE IN FORM AND IN ASPECT, WHICH SEPARATES THE TABLE OF A TWELVE-YEAR-OLD BISHOPED NIPPER FROM THE SAME PART IN THE THREE-YEAR-OLD COLT.

The difference, however, is more striking when two full rows of teeth are placed in contrast one with the other, after the manner in which they are displayed in the next engraving. In the young mouth, the incisors are arranged in a gracefully curved line; the posterior margins of the organ present little peculiarity. In the aged teeth, the prominent centre of these has retracted, while all idea of grace in the order of their disposal has departed. Each member in the old jaw evinces an inclination to become equally prominent, and the posterior borders evince an obvious angularity.

Then, if the marks in each are examined, the central cavities in the "bishoped" have jagged edges, while, from these indentations arise certain excentric lines, which invariably run towards the circumference. Such lines evidently were not made with any design. They were caused either by the inaptness of the operator, the coarseness of the tool with which he worked, or they were provoked by the natural struggles of the animal that was subjected to a merciless operation. The marks, moreover, are of a deep black colour, while the lines are remarkable for sometimes being of a lighter hue than the surface on which they repose.

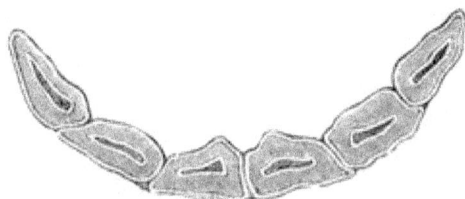

DIAGRAM, SHOWING THE TABLES OF A NATURAL ROW OF FIVE-YEAR-OLD TEETH.

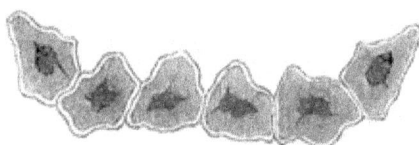

THE TABLES OF THE DISHOPED TEETH WHICH BELONG TO AN ANIMAL SIXTEEN YEARS OLD.

There are, however, other signs which faithfully denote the age of the quadruped. The permanent incisors, when first cut, are almost perpendicular, but, as years accumulate, these organs assume a more horizontal direction. The tushes, also, when they first appear in the mouth, point forward. These members, after a time, become straight; but, as age progresses, they ultimately lean decidedly outward, and at length incline backward. Besides these well-marked indications, from the disposition of the front nippers to arrange themselves in a line, only two can be seen in old quadrupeds when the mouth is viewed from the side; while the membrane covering the gums altogether loses its fleshy hue, becoming evidently thick, yellow, loose, and baggy.

Such marked signs may, by many persons, be esteemed sufficient protection; but there are yet additional characteristics, with which all who venture to purchase horses of unknown sellers should be acquainted. The general indications of senility are strongly impressed both upon man and upon horse. Though the teeth are usually appealed to, the appearance of the mouth should not be absolutely and solely regarded. A white horse is rarely young, any more than a white-haired man is, as a rule, in the possession of youth.

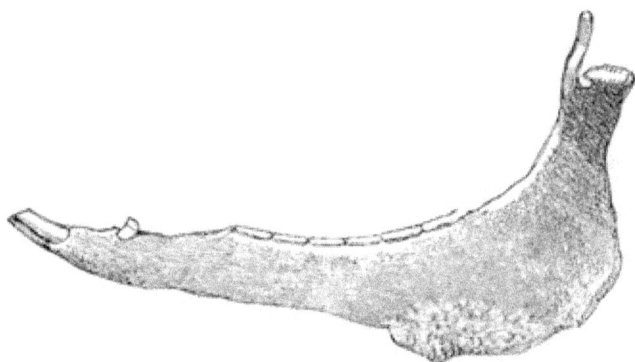

THE JAW OF A HORSE, WHICH WAS THIRTY YEARS OF AGE.

Then, as the juvenile period ceases, absorption begins to operate. Deposit no longer takes place, but, with senility, a rapid wasting ensues : both bones and flesh suffer under this new action.

The branches of the colt's lower jaw are wide apart, and in the cavity, thus formed, the tongue reposes. This space is called the "channel." The lower margins, also, of the inferior maxilla are, in the colt, full, round, and prominent. When age is present, the edges retract, the channel narrows, while

9

the lower margins of the bones appear to the fingers of the
examiner, accustomed to handle young horses, to be positively
sharp.

A YOUNG HORSE.　　　　　　AN OLD HORSE.

COMPARATIVE DIFFERENCE IN THE CHANNELS OR IN THE SPACES, BETWEEN THE BRANCHES
OF THE LOWER JAWS.

When a person, having a horse to sell, talks boastfully of
all "the marks" being present in the mouth, avoid him as a
suspicious individual. Honest men know—or at least all
honest men should by this time be aware,—that there is no
dependence to be placed in these so-called "marks"; therefore, they do not strive to direct attention towards fallacious
indications. By simply parting the lips of the animal, a judge
can see everything which he cares to behold. The kind of
teeth present are easily recognized; or, when such signs declare
the animal to be aged, the position of the teeth, the condition
of the bones, and the general aspect, enable him to guess as
to a probability. Therefore, when a gentleman requests to
see the mouth, the horse-dealer, unless specially commanded
to do so, no longer endeavours to tug the jaws asunder,—a
proceeding, which, when conducted hastily, is apt to provoke
resistance—but the groom is ordered to merely separate the
lips,—a measure to which most animals will complacently
submit.

Should the person, to whom the teeth are exhibited, by an
evident lack of recognition declare his ignorance of their
announcement, the honest dealer may slyly quiz his patron's

want of knowledge; but assuredly he will not endeavour to take advantage of it. Horse-dealers are quite as honest as—if not more honest, than traders in other commodities. There are certain blackguards, who profess to be dealers in horses, but who have no fixed place of abode, or of business. So, also, there are scamps, who style themselves travelling jewellers and itinerant booksellers; but the transactions of both classes of rogues (he whose stock-in-trade consists of a whip, or they whose most valuable possession is the mahogany box, or the specimen number which is carried from house to house), cannot be taken as evidence against the more respectable members of the calling, to which all will assume to belong. A gentleman, ignorant of any acquaintance with jockey-ship, can walk with perfect safety into the yard of any respectable dealer, look at the animals which are for sale, and walk out again, without encountering any undue solicitation to purchase. How many shops are there in London, in which a person, equally uninformed, could perform the like manœuvre?

When this is written, it is not meant to imply that a horse-dealer keeps all his stock open to public inspection. On the contrary, in most respectable yards, there are certain snuggeries, which conceal the more choice articles. The pick of these are not even open to every purchaser who can pay the price. No! horse and picture dealers are alike in one characteristic trait: each has a pride in the article he sells. The first individual will allow his dinner to grow cold, while he remains gloating over the points and beauties of some fresh acquisition. "How it would look carrying Her Majesty!" The image amuses his fancy! "What a spanker to hold a place in the Beaufort hunt!" He warms with the idea! "What a charger it would make for Cambridge at a Hyde Park Review!" He is in ecstasies at the thought! He cannot possibly decide what so much perfection is fit for. He

9 *

can never consent to treat such loveliness as a mere chattel,—
a thing to be sold and then to be enveloped in obscurity.
The animal must not be parted with to any unknown individual! The feeling common to his order forbids him to
exhibit the object of his pride to general inspection. But he
might dispose of it, even at a sacrifice, were he convinced it
would occupy such a position as he esteems it is fitted to
adorn. He then could point to the animal, and vaunt that it
came from his yard. Honour, fame, and profit must accrue
to him who could refer to such exalted dealings; therefore,
there is a strong sense of self, lurking under that, which at
first glance appears to be mere Quixotic self-denial.

At the same time, if all respectable dealers are above positive imposition, it is not every dealer who will prevent a self-
conceited novice from imposing upon himself. Such a person,
acting upon his own judgment, may be allowed to purchase
the worst screw which some yard contains, at the money that
should procure a first rate-animal. Even then the dealer has
an escape, which every form of worldly honesty will not provide. The quadruped, if not approved of, can be exchanged
within the fortnight following the transaction. To be sure,
such exchanges generally advantage only one party: but a
tradesman must live; he cannot be expected to waste hours
showing his stock, and chattering with fools for no business
purpose!

However, to protect the reader from every chance of imposition, so far as the age of the horse may be concerned, let
him attentively accompany us through the following pages;
let him, also, particularly notice the engravings with which
the text is illustrated.

A foal at birth has three molars or grinding teeth, just
through the gums, upon both sides of the upper and of the
lower jaws. The little animal, however, generally displays
no incisors or front teeth, but the gums are inflamed, and

evidently upon the eve of bursting. The molars, or grinders, are, as yet, unflattened, or have not been rendered smooth by attrition. The lower jaw, moreover, when the inferior margin is felt, appears to be very thick, blunt, and round.

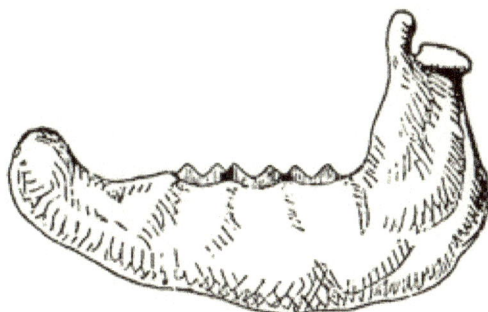

THE FOAL'S JAW AT BIRTH.

A fortnight has rarely elapsed, before the membrane ruptures, and two pairs of front, very white teeth, begin to appear in the mouth. At first, these new members look disproportionately large to their tiny abiding place ; and when contrasting with the reddened gums at their base, they have that pretty, pearly aspect, which is the common characteristic of the milk teeth in most animals. They must occasion pain

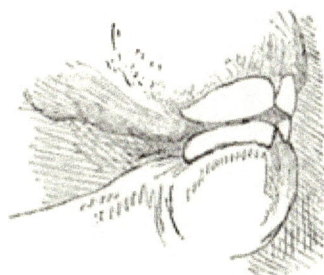

THE INCISORS, AT TWO WEEKS OLD.

to the foal at this period. The appearance of the little mouth affords sufficient evidence of that fact ; but, it is astonishing,

how meekly these beautiful creatures will submit to our
examinations of their teeth,—as though they came into the
world possessed of all confidence in man's intentions and
with every dependence upon his sympathy. Some of the
diminutive strangers seem, even, to derive pleasure from their
irritable gums being inspected. They behave, almost, as
though they recognized their future master and felt flattered
by his notice. Alas! that brutality should ever repel the
trustfulness of nature, and that experience should instruct
most of our mute fellow beings to regard mankind as enemies.

It is not until another month has passed, or until the foal is
six weeks old, that more teeth appear. By that time, much of

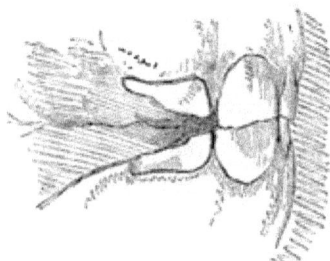

THE INCISORS AT SIX WEEKS OLD.

the swelling, present on the gums of the newly-born animal,
has softened down, though all trace of it cannot be said to
have entirely departed. The membrane, as time progresses,
will have to resign much of its scarlet hue. In the brief
period, however, which has elapsed since the former teeth
were gazed at, the growth has been such, that the sense of
very disproportionate size no longer remains. The two front
teeth are now fully up, and these appear almost of proportions
suited to the mouth which they adorn. But, when the two
pairs of lateral incisors first make their appearance, it is in
such a shape as can imply no assurance of their future form.
They resemble the corner nippers, and do not suggest the

smallest likeness to the lateral incisors, which they will ultimately become.

There is now a long pause before more teeth appear in the
mouth. The little one, in the meantime, lives chiefly upon
suction, and runs during the period of perfect happiness free
by its mother's side. Upon the completion of the first month,
seldom earlier, it may be observed to lower the head and nip
the young blades of the shooting grass. From the third
month, however, the habit becomes more frequent, until, by
the advent of the sixth month, the grinders will be worn quite
flat ; or, having lost their pointed and jagged prominences,
will, by the wear of constant mastication, have been reduced
to the state which is suited to their function.

The corner incisors come into the mouth about the ninth
month ; the four pair of nippers, which have been already
traced, being at this time fully devloped. Below is a view of
the foal's teeth, as these are exhibited at the period named.
The reader will remark, that the corner incisors, which are
depicted as through the gums, do not yet meet, though these
organs point towards each other ; neither has the membrane
of the mouth at this time entirely lost the deepened hue of
infancy.

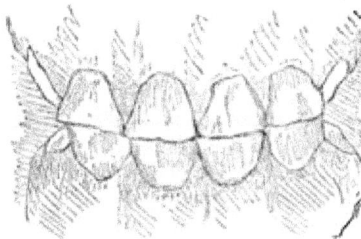

THE FRONT TEETH AT NINE MONTHS OLD.

From this date, however, the gums gradually become pale,
till, by the completion of the first year, the membrane has
nearly assumed that complexion which will endure throughout

the earlier period of existence. All the incisors are, by the first birthday, well up. The masticatory agent, although consolidated, has not, when the quadruped is one year old, entirely lost the roundness and bluntness of its inferior margin, for which the jaw at birth was peculiarly remarkable.

This fulness of the bone is caused by all the grinding teeth which are in the mouth when the foal first sees the light, being of a temporary character; the enlargement is consequent upon the jaw, therefore, having to contain and to mature the long permanent grinders which, within the substance of the bone, are growing beneath the temporary molars. To contain and to allow the large uncut teeth to become developed, before appearing above the gums, causes the small jaw of a diminutive foal to be disproportionately thick, especially when this part is compared with the same structure in an aged horse; but the mind is reconciled to its apparent clumsiness when apprised of the uses to which the organ is subservient.

At one year and three months old, the first permanent tooth appears in the head. This is the fourth molar, or that which is represented as the most backward grinder in the following woodcut. The reader will not fail to remark the greater length which the jaw-bone presents at one year old. The additional extent, also, in the opposite direction, cannot otherwise than be observed. This increase of size was necessitated to cover the increasing size of the recent molar; also, to afford room for the partial development of two other grinders, which, as age progresses, will appear behind that which is now the last tooth.

About this time, frequently at birth, little nodules of bone, without fangs, merely attached to the gums, appear in front of each row of grinders. These are vulgarly denominated "Wolves' Teeth," and were once held to be of vast importance. At present, however, they are recognized as the

simple representatives of those organs, which in other animals (as in man) render the teeth a continuous or unbroken
curve. They are, by experience, found to be harmless. It is
idle to remove these organs, especially as they generally dis-
appear with the shedding of those members, facing which they
are located.

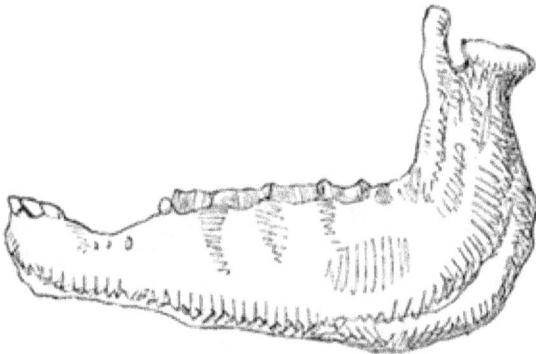

THE JAW OF A ONE-YEAR-OLD.

The changes in the teeth, after the first year of life has been
attained, are characterized by the longer periods which divide
them. Nature appears, as it were, resting to draw breath for
a mightier effort than she has hitherto undertaken. Months
have, heretofore, separated the advent of single pairs; but,
from this date, these appearances are to be reckoned by numbers and by years. The foal, to the point of its present necessities, has been provided for. It has teeth sufficient to
support and to maintain its growth.

Nature has now to render perfect the body, before the
teeth. Accordingly, between the first and the second year,
the alteration in the general aspect is very marked. All the
helplessness and pretty ungainliness of infancy disappears by
the expiration of the time mentioned. The animal's frame
then suggests something of those beautiful proportions which

it is so soon to display. Its body, however, still needs
maturing; and no one, less wanting in common sense than a
racing man, would think of subjecting the youthful and tender
form to the hardest of all actual work. The very aspect of
the creature should denote it to be unsuited for such performances. It must, to foreigners, read as strange intelligence,
that the *nobility*, who patronize the English course, applaud
the contests between two-year olds; while the *bumpkins*, who
breed horses for the general market, allow the quadruped to
enter the third year, before the colt is given over to the
breaker. Alas for the hardihood, or want of sensibility, displayed by the most exalted, when prompted by the greed of
gambling !

When about two years old, a second permanent molar,
making five grinders on both sides of the upper and lower
jaws, has broken through the gums. Preparation is also
being made for the advent of the sixth grinder, and for

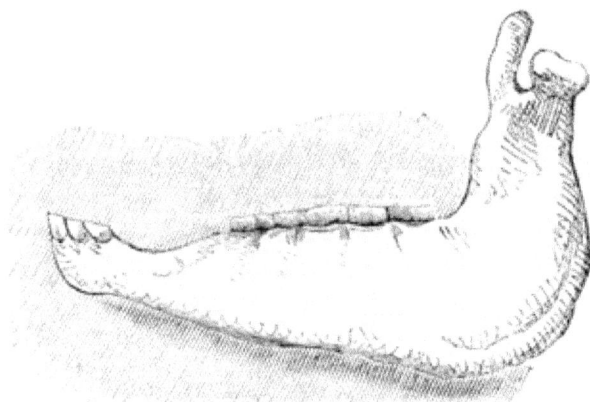

JAW AT TWO YEARS OLD.

changes in those milk molars which were in the mouth when
the animal was born. At the same time, additional width is
imperative to allow the permanent incisors to appear, when

the proper season arrives for these last organs to displace their temporary representatives.

Should the front teeth of a two-year-old mouth be examined, there will be perceived a want of that fixedness which, one

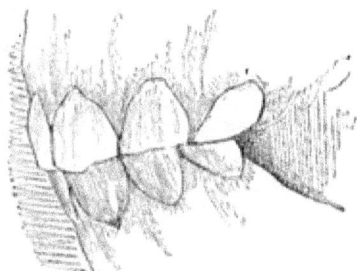

THE INCISOR TEETH AT TWO YEARS OLD.

year before, was the characteristic of these organs. The central nippers appear to have done their duty, or, at all

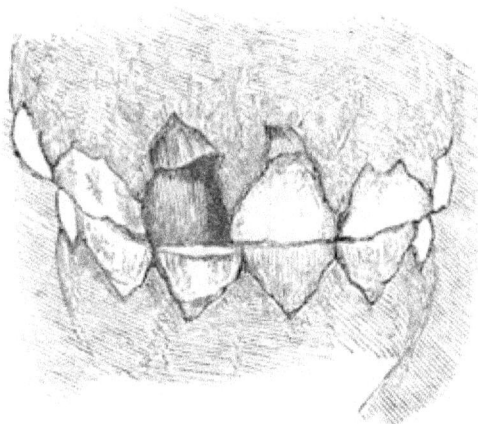

THE INCISORS DENOTE COMING THREE YEARS OLD.

events, suggest something approaching to maturity has, during the brief existence, been attained. At two years and six months old the central temporary incisors are replaced by permanent ones, which at three years old are well developed.

The age at this period ought to be absolutely ascertained, for most horses, when three years old, undergo the greatest exertion. At this period the animal generally has to suffer the instruction of a rude and an ignorant—frequently of a brutal and a savage—man, who is justly denominated "*a breaker.*" Then, should the "broken" be thought worthy of a saddle,

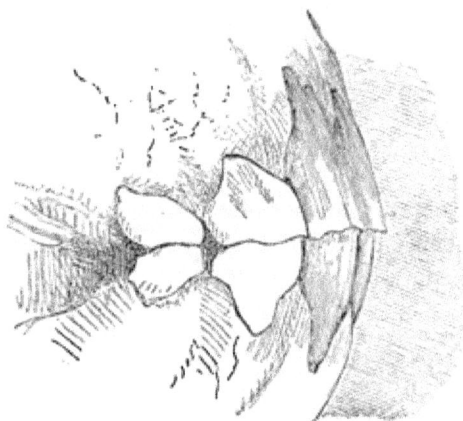

THE INCISORS DENOTE THREE YEARS OLD.

it is given up to the gentle mercies of a rough rider, and has to be tortured till it is gotten well together, and has thoroughly learnt its paces. In short, its gentle spirit has to be subdued, or, fear has to master timidity. How little does man know about that life he has been accustomed to coerce! The pride of this world prefers the compulsory drudgery of a spirit-broken slave, to the happy service of a willing friend. The horse is sent upon earth, prepared to serve and eager to share the happiness of its lord; but it is not understood; it meets with no sympathy; it is treated as a wild and ravenous

beast, whose subjugation must be enforced, and whose
obedience must be compelled.

The bit is put into its mouth when the third year has been
attained. It is driven from the field and from the cool grass;
at a period of change and of debility it is expected to display
the greatest animation, or to learn strange things from him
who teaches only with the lash or with the goad. When its gums
are inflamed; when the system is excited; when the strength
is absorbed by an almost simultaneous appearance of twelve
teeth, it is led from the plain, and made, with its bleeding
jaws, to masticate sharp oats and fibrous hay. At this age,
when fever prevails in its blood, and the growth of its frame
naturally weakens the muscles, it is expected to have leisure
to master new teachings,—animation, to show off strange
acquirements,—and stamina, to endure weight on its back;
for, should the cutting of many teeth inflame the gums and
destroy the appetite, an iron is made red-hot, and violently
forced into the mouth, under the pretence of burning away
the groom's favourite disease, " the lampas," which is purely
an imaginary disorder.

It has been described, that a three-year-old colt cuts twelve
teeth. On the next page is depicted half the lower jaw of an
animal which had seen three summers. In it, the reader will
readily recognize those organs, which are of recent appearance,
by their darker colour, by their larger size, or by their differ-
ing in shape from the other members. These new teeth are a
central incisor, and the first two grinders. The horse has two
jaws, and two sides to each jaw: therefore, the same number
being present within each side of both jaws, the teeth, already
alluded to, appear during the third year. However, even the
quantity named, rather understates, than over-rates the fact;
for, frequently, the tushes are cut during this period: should
such be the case, the colt acquires no less than sixteen teeth
in twelve months. We know what the young beings of our

own species suffer, when the gums are ruptured, and the bones absorbed by the organs of mastication; the danger, then encountered, leads to a belief, that the great agony endured is increased by a rapid growth of the body simultaneously weakening the system. The teeth are only a part of the living organism; therefore, as when a part moves, we may conclude the whole system is in motion, the advent of sixteen

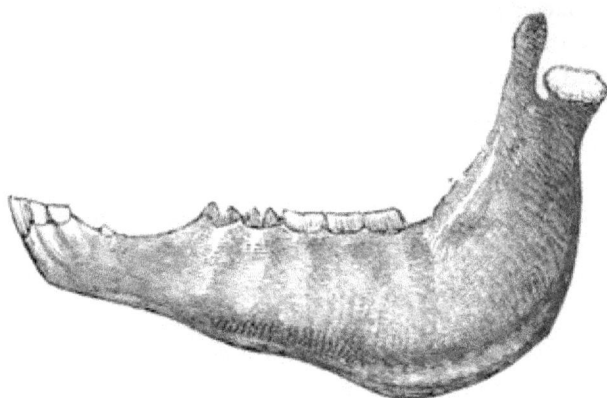

JAW OF A THREE-YEAR OLD.

huge teeth, alone, might reasonably unfit the quadruped for commencing its education, or for undergoing the severest portion of its labours. But how do the customs of humanity appear, when illumined by a consideration of the sufferings which Nature is imposing at the time the colt is tasked to its greatest exertions?

Some *very low* classes of horse proprietors will, however, make the work of the three-year-old colt as light as possible. The vulgar, generally, regard the frame at this age, as not perfectly matured; and they treat the strength as not equal to full labour. A nice practical comment is thus published upon the behaviour of those gentlemen of title and of fortune, who train, start, and make animals run races at two years old! Few members of existing society, however, will accord any indulgence to a colt during its fourth year. Yet, if the quadruped once possessed any claim upon consideration, the animal, at this period, has positive title to our forbearance. For the second effort must be more exhausting than the first; since the latter has to be accomplished with diminished power. Thus the four-year-old has to perfect as many teeth as are known to protrude into the mouth of the three-year-old.

The tushes in this view, however, must be disregarded. The precise time of appearance is uncertain with these analogues of the canine teeth in man, or of the tusks in the porcine race. They may come up at the third—they often are delayed to the fourth year, but usually appear when the horse is four years and six months old. Sometimes these teeth never pierce the membrane of the gums, it being very far from uncommon to see horses' mouths of seven years in which the tushes are absent.

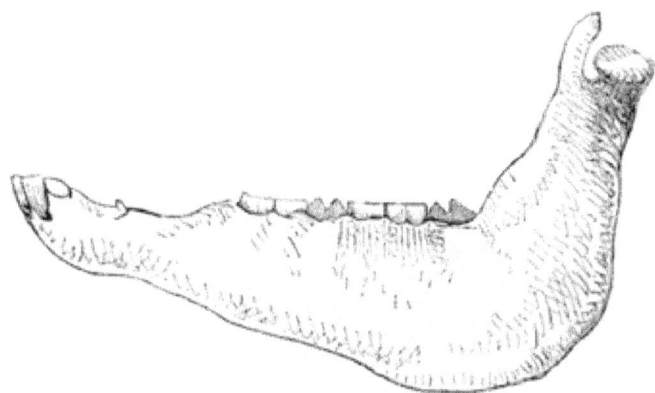

JAW OF A FOUR-YEAR-OLD.

By the completion of the fourth year, the colt has, certainly, gained twelve teeth; that is, by this time there should exist, on each side of both jaws, one new lateral incisor and two

fresh molars, being the third and the sixth in position. The
appearance of the mouth now announces the approach of
maturity; but the inferior margin of the lower bone still feels
more full and rounded than is altogether consistent with the
perfect consolidation of an osseous structure. We cannot
take cognisance of the swollen and enlarged condition of the
jaw, without being assured that some important process is
going forward within its interior. It is among the firmest
physiological truths, that Nature is a strict economist, and
never does anything without intention;—that every enlargement or every depression,—however insignificant it may
appear to human eyes,—is a permanent provision for some
appointed purpose, and has its allotted use in the animal
system. Accordingly, it is discovered, the sign we just re-

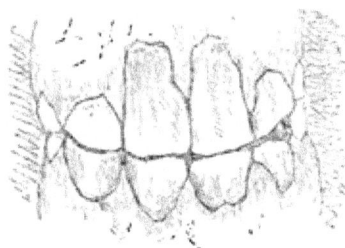

THREE-YEARS-AND-SIX-MONTHS-OLD MOUTH.

marked upon indicates the process of dentition is not finished
by the termination of the fourth year. There are more teeth
to be cut, as well as the fangs of those already in the mouth to be
made perfect. This must be a laborious effort. Nature always
toils slowly in proportion to the density of her work; when
we regard the compact structure of a horse's tooth, we may
conjecture the quantity of blood, the amount of inflammation,
and the intensity of suffering, which are necessary for its perfection. The creature, at this age, is generally urged to the

extent of his power, and is compelled to undergo severe physical exertion.

The colt with four incisors in either jaw, all fully grown, denotes that the fourth year has been reached.

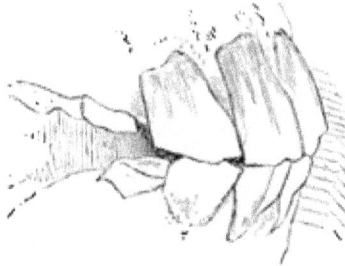

FOUR YEARS OLD.

There are still the corner milk nippers to be shed; yet while the provision necessary for that labour is taking place

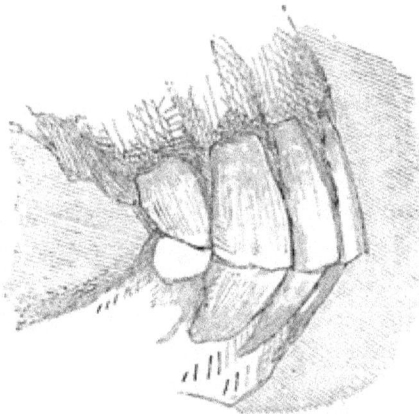

FOUR YEARS AND SIX MONTHS OLD.
ONE UPPER CORNER PERMANENT INCISOR HAS BEEN CUT; THE LOWER CORNER MILK INCISOR IS STILL RETAINED.

within the body, or while Nature is preparing her mute offspring for the coming struggle, man considers the poor quadruped as fully developed and as enjoying the prime of its existence.

10

At five years old the corner incisors are well grown, and consequently all the teeth now permanent have been cut.

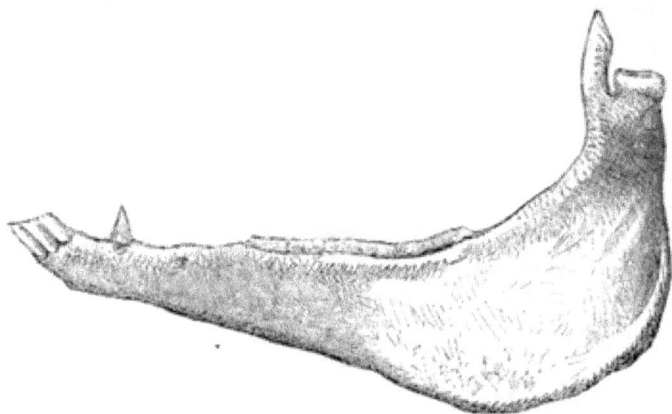

THE JAW OF A FIVE-YEAR-OLD.

MOLAR TEETH OR GRINDERS.

The molar teeth are not all of the like size, nor of one form. · The organs occupying the upper jaw are nearly, not quite, double the width of those which are located in the lower jaw. The inferior molars are the grinding agents, or the active organs of mastication. The superior teeth are simply the passive tables, upon which, or against which, the food undergoes comminution. The slab is always the lowest of the two in human mills; but nature has more to provide for than the mere pulverization of certain substances. With mastication, actually commences a very compound process. With the act of chewing, digestion begins; it was ordained, that more than any mechanical invention can accomplish, should be imperative to the due performance of this function. The benevolence of the " All Wise " instituted, that while his creatures were promoting the healthy exercise of the appro-priative necessity, they should likewise excite their enjoyment. Therefore, when pulp is masticated, the pressure of the teeth expels the juices, which fall directly on the seat of taste. When a harder substance has to be comminuted, the bulk is first shattered into fragments; the particles descending upon either side of the teeth have to be gathered up and placed again between the masticatory organs. The movements of the tongue and jaw excite the salivary glands; the broken substance becomes mingled with the secretion of the last-named bodies. Saliva extracts the savour from the food; and the tongue also brings these in contact with the seat of taste, whilst discharging its office of collecting the broken pieces.

The reader being now fully informed as to facts, may have patience sufficient to peruse an explanation of the principles on which the foregoing statements are founded. Such a mode of proceeding may, to certain methodical writers, seem to be transposing the proper arrangement. The author does not undertake to defend his actions on the score of their propriety; but he feels that he is addressing human beings, in whom a desire to know is the best possible foundation on which knowledge can be established, consequently, principles become less repulsive when communicated after incidents have kindled curiosity.

The primary molars cannot boast the length of fang, though

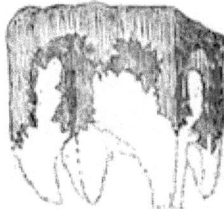

THE CONDITION IN WHICH THE TABLE OF A TEMPORARY MOLAR IS CAST FROM THE MOUTH OF A HORSE.

The dotted lines merely indicate the extent of the tooth previous to absorption.

they exhibit very nearly the same extent of superficial surface, as characterizes the succeeding teeth. They have rather

10 *

shallow roots, which are not composed of those consolidated materials that are present in their immediate successors. When the original molar is shed, the temporary tooth is not expelled entire from its position; but the pressure of the growing organ (which comes into the mouth exactly where the milk grinder stood) causes the root to be absorbed till nothing but a superficial shell has to be ejected.

The horse, in its natural state, exists on fibrous grasses; it therefore becomes essential the animal should retain the power of masticating such substances. Nature never withholds what is necessary to the well-being of her creatures. The mode in which the Common Parent provides for the preservation of this ability in the horse, is perfectly distinct from any provision that He makes for most earthly creatures.

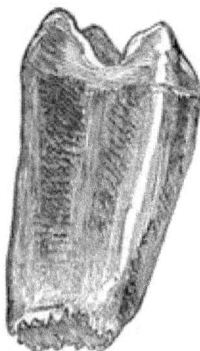

A FIRST PERMANENT MOLAR, AS IT APPEARED IN THE MOUTH, UNCOVERED.
This tooth occupied the fourth situation in the jaw; therefore, the root would require to be considerably extended by subsequent growth.

The temporary remains of a molar tooth are not shed till another organ is in the mouth at hand to permanently supply its place. But the permanent tooth does not appear ready flattened and prepared to discharge its office. It is cut with certain angular prominences upon its masticating surface, which must render the animal disinclined to employ it on the instant of its development. This disinclination allows a

panse during which the various structures can be consolidated, and, at the end of such brief space, the prominences have become blunted, while the organ, being firmly
planted, is then ready for mastication. Is it not surprising
how a plain statement of facts can reasonably account for
that disinclination to feed, which, to the groom's mind,
announces a state of disease, that shall necessitate the employment of burning iron to eradicate what the man styles
" Lampas."

There remains, however, to account for that width and
depth of jaw, by which the head of the youthful horse is
distinguished. The reader is requested to attentively inspect
the last illustration. The size and length of fang cannot fail
to awaken his surprise. Nevertheless, if this part be
regarded, it will be seen depicted as of a ragged, incomplete,
and apparently of a hollow condition. So, when the tooth
has displaced the temporary molar, and has taken its station
within the mouth, it has still to grow. The protruded portion
may be consolidated; but the unfinished extremity is denominated the cavity of the pulp. That pulp consists of a fine
bladder, on which ramify numerous blood-vessels; but the
interior of which contains simply a clear fluid. This is the
secreting membrane of the tooth. Out of this watery bag,
the wonderful chemistry of nature can extract the most con-
densed material that resides within the strong body of a horse.

Another feature of the above tooth, because it balks expectation, can hardly fail to attract notice. The dark hue of the
outward covering, being abhorrent to human notions of youth
or of purity, is generally attributed to dirt. The tooth of the
horse is, however, composed of three substances. A tough
and fibrous material, called crusta petrosa; a thin layer of
crystalline deposit, named enamel; and a kind of compact
bone, spoken of as dentine. They occur according to the
order in which they are named. The bone exhibits a yellow

tinge, and is present in the greatest quantity, for it forms the
inner bulk of the tooth. The crusta petrosa is a comparatively
thick external envelope, being about five times the thickness
of the enamel, to which it is an outward protection. The
components are thrown into various convolutions; but the

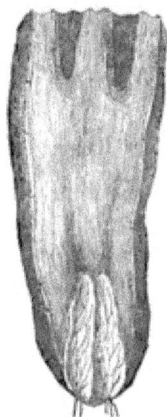

SECTION OF A MOLAR TOOTH.

order alluded to is always preserved. The bone or dentine is
invariably the internal substance; it needs to occupy such a
position, as within it the sensation resides. The crusta petrosa and the enamel may be tampered with without perception being aroused; but the dentine is capable of communicating the acutest agony; and it is upon the dentine that
rogues operate, when they "bishop" an old horse.

To convince the reader that Nature has not needlessly sacrificed the whiteness of the horse's tooth, the author will dilate
fully upon the many services afforded by the dark-coloured
crusta petrosa. To render the explanation more intelligible,
reference will be here made to a common tool seen every day
in the hands of an ordinary mechanic. The bricklayer's
trowel appears to be nothing more than a thick layer of metal
but it is hourly put to uses for which iron would be too soft,

and steel would be too brittle. Therefore, the blade is composed of a thin layer of steel, inclosed within two comparatively thick layers of iron. By the combination of opposite qualities, perfect utility is produced; and this trowel, it seems hard to believe, was not suggested by that arrangement which is conspicuous in the horse's tooth.

The enamel, hard, brittle, and readily fractured, but presenting a fine or a cutting edge, is developed as a thin layer, convoluted upon the sides of the dentine, and securely covered by crusta petrosa. That the incisive substance may fulfil its office, may sever or comminute the tough and fibrous herbs, upon which the equine race subsist, it is inclosed between two elastic bodies, the whole being held together by the vessels, which pass from the exterior to the interior of the organ, though these vessels do no more than travel through the enamel, without nourishing or supporting it; the latter structure being of a crystalline nature, or strictly inorganic, therefore not fitted to appropriate nutriment.

The crusta petrosa is, however, of further use than has been already stated. The horse's grinders are generally supposed to be gifted with a power of growth, whereby they are enabled to repair that perpetual loss of substance to which their employment must subject them. The teeth, certainly, are not perfected when the crowns first appear in the mouth; so far the opinion is capable of being upheld. But, when once completed, the dentine is not endowed with any innate ability to renew its loss of substance. The wear consequent upon continual use, is provided for by the length of fang, which characterizes the permanent molar of the quadruped. As the surface gradually decreases, so are the lower parts of the teeth, by the contraction of the jaw-bones, forced into the mouth, while the outward investing substance—the crusta petrosa—being gifted with a limited power of increase, is enabled thereby to firmly retain the protruded fang in its new position; although the contraction of the bones, which is always going forward as age advances, does not necessitate the power of growth should in early life be largely exhibited.

With almost every form of being, as years accumulate, the ability to masticate becomes enfeebled. It is with the horse as it is with other animals. The thin coating of enamel does not extend to the ultimate root of the fang; so that, in advanced age, the power of the molars is almost destroyed by the absence of the cutting agent upon the grinding surface. The chief component, moreover, or the dentine, diminishes in quantity as in solidity; the last portions of the molar, therefore, could not fill the socket, only for that ability to increase, with which the crusta petrosa is gifted. Upon the extreme roots of the grinders, taken from the jaws of very old horses, this substance is always found in great abundance. In illustration of this fact, a sketch made from the tooth of an aged quadruped, is inserted below; the body has been sawn

SECTION OF AN AGED MOLAR TOOTH.

asunder, to exhibit the proportions and the substances that entered into its composition. The reader will remark certain dark lines upon the dentine. These indicate the places where existed the cavity of the pulp, which once served to nourish the organ; but it is lost as vitality lessens with the advance of senility. Does not the reader, as he inspects the engraving, perceive the wickedness and the folly of placing harsh and dried food before a creature which nature, in age, deprives of ability to comminute such a form of sustenance?

The permanent incisors are not cut after the same manner as the molars. The nippers being merely employed to bite the grass, a wide vacancy does not necessarily incapacitate the other portions of the excising apparatus. A blade can cut, even though a large notch exist upon its edge. Whereas the points which are developed upon the upper surfaces of the newly-cut molars, must render the grinders entirely useless; although the short period of enforced abstinence, which announces the appearance of a fresh double tooth, may be Nature's own medicine to quiet a feverish system, burning with morbid excitement.

The front milk teeth have fangs when they appear in the mouth; but no fang exists when the primary members are shed. The root of the temporary organ, when perfect, however, resembles that of the permanent incisor. It is only

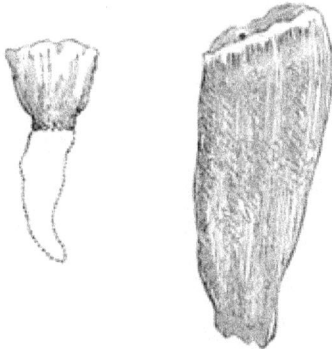

A MILK AND A PERMANENT INCISOR TOOTH.

sufficient to fit the member for its purposes. In the same canal as was occupied by the milk tooth, the permanent incisor generally appears. Much suffering must attend the absorption of bone; yet during the time the huge permanent nipper is forcing its way through the narrow channel, which held firmly the diminutive milk tooth; and while the smaller fang is by pressure being also absorbed, the colt receives no consideration at the hands of the groom, or of its master. Both are equally ignorant of the necessity for kindness; but each regards any indication of pain as one of those visitations of disease, to which young horses are said to be peculiarly liable.

When the foal has shed the front milk teeth, the entire of the serviceable or visible portion of these members is displaced. They are endowed with no power to supply any diminution of their substance, neither are they capable of renewal; whereas the long permanent incisor may be viewed as all tooth, and possessing no fang; for, as the upper portion wears, so does the lower part protrude or supply its place. The two teeth, however, present a strong contrast, when considered as organs, both occupying one cavity, and both united to fulfil the like uses, in the same animal. The illustration, last displayed, represented a permanent and a temporary incisor; the uneven mark, dividing the milk tooth, indicates the appearance of the organ after the absorption of the fang causes it to be cast from the mouth, while the dotted line shows the shape and the extent of the fang previous to its absorption by pressure.

The amount of root natural to the permanent incisors, enables those organs, as years increase, to alter their arrangement, length, and direction, without being displaced. In youth, the united front teeth compose a curve, or almost a semicircle. In age, the same members incline toward a straight, or at best, form an irregular line. In the colt, the teeth are flat, smooth, and filbert-shaped; but in the old animal, they are decidedly long and angular. When the permanent teeth first appear, they are nearly perpendicular; but when they have been a long time exposed, they protrude almost in the horizontal direction. Looking from the side at a young mouth, the spectator can behold half the nippers; but when inspecting the old teeth from the same point of view, two only will be visible, though the full number shall be present in the mouth. In the aged quadruped, moreover, the narrowing of the incisors allows the spaces between the organs to be vacant. Within these spaces the food accumulates, which, being there retained and becoming black, looks as though the creature had been chewing tobacco. Such signs are too fixed to be disguised. The accumulation of blackened food, it is true, may

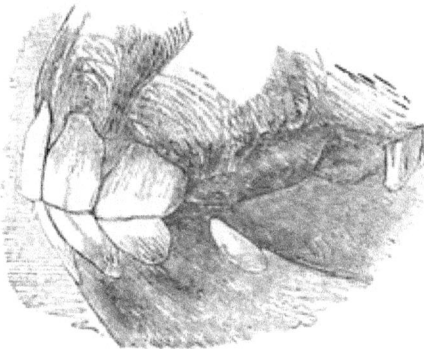

SIX YEARS OLD.

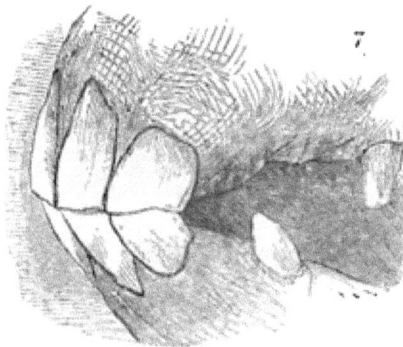

SEVEN YEARS OLD.

be taken away; but its removal will leave the interspaces, if possible, still more conspicuous. So, also, the long teeth may be shortened; but they will not be elevated to the perpen-

dicular, or changed to a filbert form, or restored to the semicircular arrangement.

The tushes likewise may be regarded. These teeth are sometimes absent in mares, and, in animals of the female sex, are

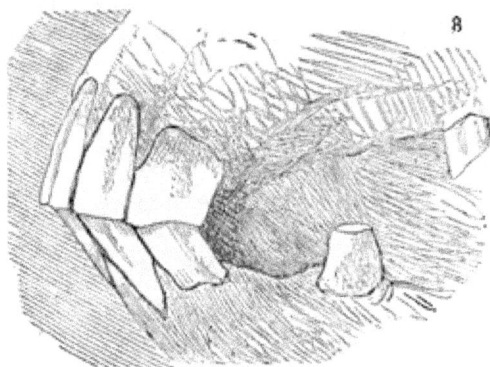

EIGHT YEARS OLD.

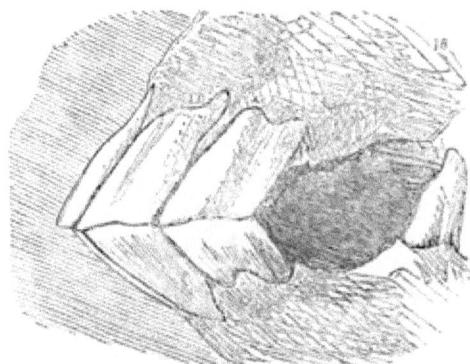

TWELVE YEARS OLD.

THE INCISORS OF HORSES OF DIFFERENT PERIODS OF AGE AFTER THE FIFTH YEAR.

seldom developed of the size which they commonly exhibit in
the male. When first cut, the tush is spear-shaped, having
well-defined grooves running down its margins. As age
advances, all pretension to this form is lost. The tooth either

becomes very flat upon its crown or it may be rendered level
with the gum; else it grows very long, looking more like a
coarse spike, than the organ it really is. Also, when it originally appears in the young mouth, the tush ranges evenly with
the parts from which it grows, and points forward. As senility is attained, the member is directed outward; with extreme
old age, it faces backward. The contraction of the jaw causes
the tongue to protrude from the free spaces between the teeth,
while the consequent shallowness of the canal formed by the

TWENTY YEARS OLD.

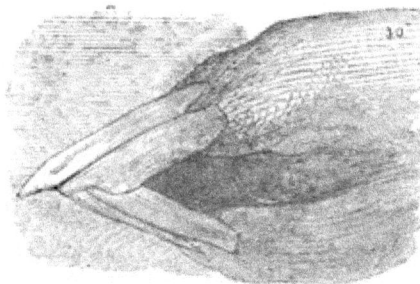

THIRTY YEARS OLD.

THE INCISORS OF HORSES OF DIFFERENT PERIODS OF AGE AFTER THE FIFTH YEAR.

branches of the bone occasions the saliva to dribble forth
when the lips are parted.

The indications of extreme age are always present, and though during a period of senility, the teeth cannot be literally construed, nevertheless, it should be impossible to look upon the "venerable steed" as an animal in its colthood.

No man can, accurately, interpret the signs of the teeth after the fifth year. A guess, more or less correct, can be hazarded; but nothing like confident judgment can be pronounced, subseqnent to the period just named. Cases will frequently occur which shall set our best endeavours to be correct at defiance. But for such instances it is not difficult to account.

The difference between the times of birth in various animals, it is true, may cause different aspects in the teeth, and even induce men, in obedience to the rules of the Jockey Club, to call a colt four, which truth and the teeth declare to be only three. Horses may therefore be readily reckoned as older than they really are; but there is a general belief, that rogues in Yorkshire can make the teeth say five, when the actual age is only four; or in other words, can so successfully tamper with the mouth, as to induce the teeth to belie the actual age. Ignorant people have a blind faith in the power of those who chance to be more knowing than themselves; but the reviser can only regard the general belief in "Yorkshire fives," as illustrating the total unacquaintance of the public with all that concerns equine economy.

So particular people appear to credit Yorkshire horse-dealers with an ability to perform " such tricks." No doubt they have every wish; but the reviser questions whether they have yet attained the power to compel Nature at their bidding. All they are said to do, as pulling out the milk teeth, firing and blistering the gums, are like the arts which were formerly used to raise the evil one; and, in the writer's opinion, about as likely to be attended with success. Cruelty is more calculated to retard than to promote development. However, if the

mouth exhibit the signs proper to a five-year-old, the animal
may with safety be purchased, as being of that age.   Should
it be younger than five, the owner is the gainer; since the
teeth do no more than indicate the development of the body,
and an early maturity is the best evidence that the quadruped,
during the previous years, has been tenderly nurtured.

As age advances the teeth naturally wear down, but the
mode, and kind of food upon which the horse is fed, operate
largely in permitting or controlling inordinate growth.   The
stable diet throws the incisors out of use.   These teeth, in the
domesticated animal, are employed only to grasp a little hay
and to pull it from the rack.   They are of no further service.
One of their popular names, "nippers," is in general a misnomer, for they are permitted to nip nothing ; much less are
they allowed to exercise their incisive faculty.   Therefore,
being thrown out of use, the members have no function to
control their natural growth.   They continue to protrude as
age advances, till, by the thirtieth year, or by the time the
quadruped has attained half the period of its natural existence,
the front teeth have become long spikes, and are actual deformities within the mouth they were designed to adorn.

CHAPTER V.

THE food grows abundantly on the surface of the earth; every fresh mouthful necessitates an additional step; for the animal, when free, walks as it eats, and lowers the head to collect its sustenance from the ground. Mankind imprisons the poor life; the hay is placed level with the ears, and the corn is given even with the chest of the animal. Nay, the very groom, when he permits water to be imbibed, raises the pail, resting its edge upon his knee. Nature enabled the horse to feed by night,—when the air is cool; when all is quiet; when the grass is moist, and when the flies are not abroad; then, the emblem of concord pastures in peacefulness.

No doubt the natural food consists of the plants and grass which grow on the earth's surface. The horse does not graze without selection. Certain herbs are scrupulously avoided; others are eagerly sought for. The animal does not eat straight before it; but the head moves to either side, each mouthful being carefully collected with the lips, before the juicy tops of the plants are operated upon by the teeth. The horse feeds only off the growing ends of the grasses. The varying herbage may be supposed to present numerous savours to the keen scent of the pasturer, and a fresh flavour may be relished with each new mouthful. Nature has evidently scattered variety, where the dull sense of man can perceive only sameness; and, to the temperate palate of a horse, the verdure of the fields may afford a delicious and an ever-varying banquet.

The instinct, which enables the animal to make a selection among numberless growing plants, fades and is lost when moisture has departed with the colour, and the perfume natural to the herbage has been changed by art. The animal perceptions may be puzzled, for art can defeat instinct. Some quadrupeds, as if much perplexed, will pick the hay, eating little; but spoiling more than is consumed. Others appear to distaste the preparation, and these refuse it altogether. Few inhabitants of the stable will accept all that may be placed before them; though the rejection may depend more upon the fastidiousness, begotten by captivity, than be generated by positive dislike. Few animals exhibit either choice or discretion in the selection of certain portions of prepared fodder. The rejection of particular parts seems to be guided only by fancy or caprice. That which in the green state would be abhorred, when "cut and dried" may by preference be devoured.

But although such is the case, during the cold months of the year it would be impossible for us to feed our horses sufficiently unless we dried our grass in summer, in order to store it up, as hay, for winter consumption.

Few people can tell a good from a bad sample of hay. Vast quantities of that, which no proprietor should oblige his imprisoned slaves to consume, are daily sold; some persons even prefer particular kinds of produce; whilst others, urged by parsimony, will purchase only damaged hay. There should be, however, in this substance, little room for the exercise of choice or of discretion. The characteristics of good hay are very marked, and such only should be purchased by the careful horse-owner.

It is the intention of the author to offer some remarks upon this simple, but excessively important topic. The comments

11

will be accompanied with tinted wood engravings, which will
help the judgment, though these cannot inform the reader on
every particular. Therefore, he must kindly assist the writer,
as few things are more difficult to describe than taste or smell ;
since these senses are always under the control of individual
predilection.

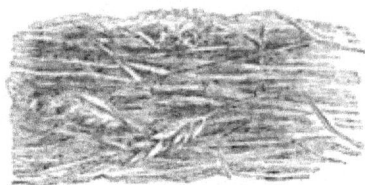

THE CHARACTER AND THE COLOUR OF UPLAND HAY.

Upland Hay should look clean. Every fibre should appear
distinct. The colour should be bright and should convey an
idea of newness. No dust ought to be present; neither should
the sample, however much it may have been disturbed, lose
its prominent features. The constituents will all point pretty
much in one direction. Of course this order is not so absolute as to appear like arrangement; but the confusion, which
generally marks the fibres of the after-meath, is never present
in a fair sample of well-carried " Upland hay." The scent
is commonly very pleasant—not so strong as—but in other
respects little different from the perfume of new-mown hay :
to most people its odour is highly agreeable. Weeds should
not be abundant ; but the presence of foreign growths is
clearly indicated by their darker hue, by the browner tint, and
the fuller form. The stems should not have shed the seeds,
though grasses vary so much in the period of their ripening,
that it is vain to expect some will not have broken this rule.
When a portion is placed within the mouth and is masticated,
it rather communicates a mild and pleasant flavour, than yields
a strong or pungent taste. In short, cleanness and delicacy

are the prominent characteristics of "Upland hay;" which some growers imagine is scarcely injured by long keeping. New hay is certainly objectionable. But the year's growth is wholesome feed by November; and, in the author's judgment, it is best when it first comes into use.

THE CHARACTER AND THE COLOUR OF LOWLAND HAY.

Lowland Hay. This kind of preserved grass lacks the bright colour, being more tawny than the preceding; indeed, the absence of the green tint is conspicuous, and can hardly fail to be remarked. The arrangement of the fibres is not so well preserved, neither is the crispness or the newness of aspect, for which "Upland hay" is notable, to be remarked in the "Lowland truss." The flowering heads to the stems are all but absent. When felt, it communicates a sense of softness. If rattled, no brisk sound is elicited. It has a stronger and a more pungent perfume. The odour is very far from being so delicate; neither is the taste characterized by any pleasantness of flavour. When placed between the teeth, mastication communicates a sense of softness and toughness: the taste is coarse, almost disagreeable: at first it is vapid, though after a short space, a certain amount of pungency is developed. The woolly texture; the want of boldness in the component parts; their comparative smallness; with the washed-out aspect of the whole, and the confusion of the mass, should prevent a novice, even, from accepting "Lowland" for "Upland hay."

11 *

Rowen, or After-meath, presents a greater confusion than even "Lowland hay." The softness is more conspicuous; flowering heads are only occasionally met with; the stems are few in number, are small in point of size, and form no prominent feature of the whole. This species of fodder lacks perfume altogether; but, as regards colour, it may have a slight greenish tint clinging faintly to it. Still, by its want of the brisk or the healthy aspect, and by its darker hue, it is at once recognised for the thing it is,—an unseasonable produce, reaped late in the year, and got up long after the

A VERY FAIR SAMPLE OF ROWEN OR OF AFTER-MEATH.

freshness of spring had departed. To the mouth it imparts a strong and slightly bitter taste. The odour is not objectionable, although it does not approach to a perfume. Horses, which have been accustomed to the better sort, refuse Rowen, or only accept it after actual hunger has been experienced.

Clover Hay is universally mixed with grass and weeds. A good sample of this produce, a novice might easily reject as being too foul a specimen for his approval, and the hay of the second crop (which is not generally remarkable in that particular) be selected in preference. The stems also appear to bear a large proportion to the whole, when compared with the flowers and the leaves. The fact of the stalks being rarely viewed in the clover-field may render this feature the more conspicuous. But the stems are hollow, and consequently lose little bulk when dried. The flowers and leaves,

on the contrary, are juicy; and no insignificant portion of
their substance is, apparently, lost during evaporation. In
the first cut of clover, however, the stems, though numerous,
are comparatively fine, and the leaves, though dark, have no

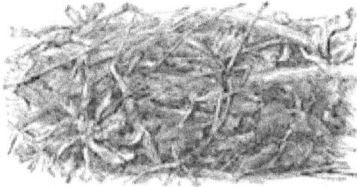

A SPECIMEN OF THE FIRST CROP OF CLOVER HAY.

tinge of blackness. The flowers are abundant, and faded, of
course; but they still retain indications of their original
colour. Though compressed, they nevertheless suggest what
has once been their figure. In taste, a marked resemblance
is recognised between the slight flavour of the hay and the
strong aroma of the growing plant.

The Second Crop of Clover is distinguished by the grasses
and weeds of the first cut being all but absent. The stems
are larger, firmer, and bear a greater proportion to the
whole. The flowers are not so numerous, and are more dingy
in appearance, as well as apparently less carefully preserved.

THE SECOND CROP, OR AFTER-MATH OF CLOVER.

Mastication, also, enables to be recognised a coarser and a
stronger flavour than characterises good hay of the spring's
harvest. The leaves approach near to a black tint. When a

truss of the first and one of the second crop of clover are placed together, the last appears remarkable for depth of colour.

Heated, or Mow-burnt, Hay is that which has been subjected to such uncontrolled fermentation as shall scorch the substance, and, if not checked, would ultimately fire the stack. A certain amount of fermentation is needful for the development of sound hay; but should the necessary action be suffered to proceed too far, "heated, or mow-burnt, hay" is the result. Most horses will eat this kind of fodder with appetite when it is first presented; but after the novelty of the diet has subsided, there are few animals which do not apparently loathe

A VERY EXCELLENT SAMPLE OF HEATED HAY.

such produce. The illustration by no means represents the worst specimen which the author has encountered, but it is of that medium character which best conveys a just idea of a general subject. From this sample, however, certain leaves could be chosen that are perfectly black, and which, when attempted to be rolled between the fingers, would crumble into powder. Such a peculiarity, together with the darkened hue, affords the easiest means of recognising this provender, which, although some silly people, by preference, employ in their stables, is very far from being a wholesome food for horses. Burnt vegetable matter produces potash; therefore there can be no cause to reject, as a groundless prejudice, the assertion that much "mow-burnt hay" will occasion diabetes.

It has a powerful odour, resembling the mixed smell which pervades a public hay-market; but the taste has little to distinguish it, being somewhat vapid.

Weather-beaten Hay is equally devoid of smell or of taste. It has a ragged, a confused, and a broken aspect. The hue is deepened, but the colour greatly depends upon the period of its exposure; the soil on which it has lain; the amount of wet to which it has been subjected; and the condition in which it has been " got up." So delicate a produce as carefully prepared hay, of course, cannot be long exposed to the

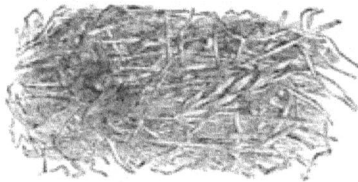

WEATHER-BEATEN HAY.

effects of wind and rain without its more choice qualities being deteriorated, while to the extent of its deterioration, of course, the farmer can oppose no check. Therefore, a fair general specimen, exhibiting the common characteristics of the majority of samples, is submitted to the reader; but it cannot be expected that a single illustration should embody the multiform aspects which are generated by diverse and powerful influences acting upon a perishable substance.

Musty Hay is readily recognized by its strong and peculiar smell, resembling the refuse which has been employed to stuff articles of cheap furniture. This it likewise calls to mind by its rumpled and confused appearance. It should never be offered to any animal as a substitute, even, for better food.

" Upland Hay," as will be seen by the foregoing remarks, is a fair general fodder for the horse. To it, however, a portion of clover hay should be added; but this last is best given in the form of chaff. Ready-cut chaff should never be purchased, because most persons have extraordinary notions as to the ingredients suited for such a form of provender. Hay, which the animal refuses to touch when placed in the rack, is often salted and cut into chaff. Thus seasoned, and, in such a shape, being mixed with corn, it may be eaten. The horse is imposed upon by the salt and the oats which were mingled with the trash; but the sane proprietor has only to calmly inquire of himself—whether that savour which disguises the taste can also change an unwholesome substance into a wholesome nutriment?

It is likewise a prevailing custom to cut straws of different kinds and to throw the rubbish into the chaff-bin. Such a practice is spoken of as among the improvements of modern horse-feeding. The quadruped may consume this species of refuse, but it is, in the author's judgment, not a matter for doubt whether such articles merely distend the stomach or whether they can nourish the body. People who advocate cheapness may be favourable to the use of straw; but these persons should not deceive themselves, far less ought they to impose upon others, by asserting so exhausted a material can possibly prove a supporting constituent of diet.

Within the stem of the ripened wheat plant no sap circulates. All the strength of the growth has gone to the seed. Were not the sapless stalk cut and preserved by man, it would shortly topple over, and, by decay, be mingled with the soil. It is well understood that grass, after it has shed its seed, is unsuited for making a nutritious hay. Grain-yielding plants are only cultivated grasses; and the art which has enlarged the seed and lengthened the stem cannot pretend also to have mastered the laws of Nature by having endowed a refuse material with nutritious properties. Persons who desire to have straw mingled with the manger food of the horse, should take some pains to procure articles rightfully prepared. The plants should be mown while green; be properly treated, stacked and husbanded with more than the care usually bestowed on ordinary stems. The same rule should be observed with regard to bean-stalks, or whatever else is to be severed into lengths, and is to be esteemed a fitting food for the horse.

Thus prepared, the wheat stem might prove worthy the repute which is, at present, bestowed upon its exhausted representative. When harvested after this plan, the stalk would retain all that virtue which, at a later season, is expended upon the seed. It would nourish as well as distend. Indeed, the popular custom of giving horses that for food which adds to the bulk of provender, but does not support the system, cannot be too strongly reprobated; yet such a practice is followed in the great majority of existing stables. The animals, to satisfy the cravings of appetite, are compelled to devour more than their diminutive stomachs should contain. Over-gorging is likewise promoted by the habit of subjecting all kinds of horses to prolonged and unnatural periods of abstinence. The consequences of such customs are exemplified in the attenuated stomachs of most old subjects. Often this viscus, upon the muscular and secretive actions of which the health and the strength are dependent, when taken from the body of an animal which has long been subjected to the abuses practised in the modern stables, is of so stretched a nature, as to be semi-transparent, and sometimes as thin as brown paper.

When a horse returns home, after a long fast, it is most unwise to place the famished life before a heaped manger. First attend to its immediate requirements. These satisfied, and the harness removed, a pail of gruel should be offered to the animal. The writer knows it is said by many grooms that their horses will not drink gruel, and few will the mess that is usually presented to them, which consists of a little oatmeal mixed with hot water; and this, under the name of gruel, is offered to the palate of a horse which long abstinence has rendered fastidious.

Some horses, however, purge, when brought home after a long fast. Such animals are generally of a loose and weakly constitution. For creatures of this description the bran would prove injurious, and an additional pint of meal had better be boiled in a quart of water, which, when mixed with the solid from which the gruel has been strained, will constitute a moist and highly nutritious diet for a delicate horse. The author has, for experiment, tried this form of food upon several quadrupeds, which he was assured abhorred everything like mash or gruel; but only in one instance was the preparation not eagerly consumed. In the exceptional case it was not entirely rejected, being partly eaten; but the writer suspects the apparently dainty quadruped had been previously supplied with a more than usual quantity of oats, as the behaviour rather testified to want of appetite, than denoted any positive dislike of the nourishment which was before the animal.

Besides hay, corn is commonly used in this country as a food for horses. The corn of the English stable is almost confined to oats. In foreign lands various substances are employed. General, however, as the adoption of oats may be in this kingdom, few, very few, persons, beyond the limits of the corn-market, have any distinct notion concerning this kind of grain. With the vast majority an oat is an oat, and all oats are of one kind. In exceptional cases, gentlemen are partial to oats of some particular hue. Certain persons will purchase only a black oat; another class prefer a full golden tint, to distinguish the kind they favour; while a few admire a whiteness of husk. Such differences, however, do not affect the grain; the colours are limited to the chaff,—the kernels of all are of one tint.

The kernel, or the mealy substance of oats, differs in each variety of corn. One sample shall be thick in the husk, and possessed of a superabundance of beard; but the body of such corn will be narrow, also of contrasting sizes and of various colours.

The inferior specimens are commonly mixed with other seeds, with pieces of stick and portions of straw, as well as sometimes adulterated by the presence of other grain. These oats may

KONISBERG OATS.

PETERSBURG OATS.

impart a saltish flavour to the taste; likewise they may have a faint, smoky, or fusty odour. Such corn seldom weighs more than twenty-two pounds to the bushel.

Another sample, of a different country, will rattle briskly as it is poured from the bulk into the palm. Such has a clean aspect and almost a metallic lustre. It is full or plump, being positively beardless, and exhibits no more husk than is needed to surround the kernel of such grain. When attentively scrutinized, perhaps, no specimen of oats will be found to be all of one size; but no very striking inequalities will catch the attention, when the better sort are viewed. These are entirely tasteless; and do not even suggest the possibility of a scent appertaining to them. Corn of this quality is too valuable not to be carefully harvested; consequently, the hardest pressure of the thumb nail leaves no indentation; while the kernel rather chips than tears asunder, when compressed between the teeth.

The absence of beard, however, is not an invariable sign of excellence: if the weight per bushel be heavy, this feature should not be too strongly insisted upon. Some good corn is distinguished by a greater length of husk than is requisite simply to surround the kernel; but such atones for this peculiarity by the bulk of the grain. It is true, that a sample of this kind seldom attains to the highest weight, and the purchaser loses somewhat by an excess of chaff.

OATS.   KERNELS.   OATS.   KERNELS.

BEST SCOTCH OATS.   ENGLISH OATS, FROM CANADIAN SEED.

Yet, in England, which country on the continent is esteemed to be a land of horses, very few stables are supplied with other

OATS.   KERNELS.   OATS.   KERNELS.

NEW IRISH FEED OATS.   FIRST-CLASS SWEDES.

grain than that of an inferior description. The better kind is bought by the miller and the trainer of racers or hunters.

The inferiority of most corn, however, seems not to disturb domestic tranquillity. The majority of proprietors open an account with some neighbouring chandler, and the groom is empowered to fetch the provender, which the horses are supposed to consume. Dealers in grain do not enjoy unsullied reputations. It is a custom with grooms to exact ten or five per cent. on all the master's bills which refer to the stable. The gentleman, therefore, always purchases his fodder very dearly, where such an arrangement exists.

Oats should never be bought by measurement, but should invariably be purchased by weight. A prime sample will weigh forty-eight pounds to the bushel; whereas the author has heard of, although he does not pretend to have seen, oats so very light, that the same bulk was only equivalent to sixteen pounds. However, a grain, which is professed merely to reach twenty-two pounds, is to be met with in every market.

The difference of weight should be. more than accompanied by an equivalent diminution of price; because a prime oat of forty-eight pounds will yield thirty-six-pounds of pure grain,

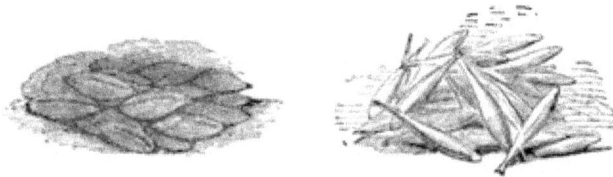

HEAVY AND LIGHT OATS AS EACH LIES IN THE MEASURE.

after the chaff has been removed. A fair oat gives half its weight of kernel; but an excellent sample will afford three-quarters of its entire weight in prime nutritious substance; whereas a poor specimen will produce no more than eight pounds of clean corn to the bushel measure!

Consequently, supposing a choice sample to sell for thirty-six shillings, the inferior article can be worth only eight shil-

lings the quarter; for no man can esteem the husk as a food
suitable for any living creature, nor would any person purchase
such utter refuse, even at the fraction of a penny per pound.
Cheapness in such particulars is, therefore, very far from the
truest economy.

The animal is doubly defrauded where poor corn is served
out by measure. The grain, in the first place, contains less
nourishment: in the next place, the solid bulk is not the same;
because the husks not only occupy more space, for, by acting
as props to one another, frequently clear cavities are formed.
Therefore, were the light and the heavy corns required to fill
a given measure, to be counted, probably no vast difference
would be discovered in their number. The reader must, however, himself determine how far it is possible for a horse to be
cheated, without the master suffering from the fraud in its
effect.

Further injury is inflicted by permitting the quadruped to
consume only an inferior corn. Whoever will carefully
examine the drawings given of oats in the present division of
the book, can hardly fail to remark, that the denuded kernels
appear of a size disproportioned to that represented as the
dimensions of the perfect grain. The microscope makes plain
the source of this apparent disparity. The epidermis, or the
covering of the kernel, is coated with numerous fine hairs,
which are too small to be perceptible to the unaided vision.
These hairs are closely compressed, when surrounded by the
natural envelope; but when released from the husk the hairs
expand, and thus occasion the naked eye to behold something far too large for the case from which it has recently
been released.

In the inferior sorts, the hairs are rather longer, and likewise more numerous, than in the better kind of corn; while,
of course, the covering, according to the smallness of the
grain, becomes serious, when regarded as a proportionate

weight of the whole. These diminutive hairs are perfectly
indigestible and entirely indestructible, when taken into the
stomach. The peristaltic action releases them from the surface
of the kernel: being set free, they are frequently felted together
by the moisture and rolling motion of the stomach. However small the hairs may be separately, nevertheless, by their

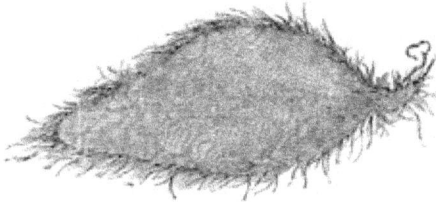

A MAGNIFIED ENGLISH OAT.

union they form masses of immense size; provoking such
serious impactment as often leads to a terrible and a fatal
issue. A further reason, therefore, exists for employing good

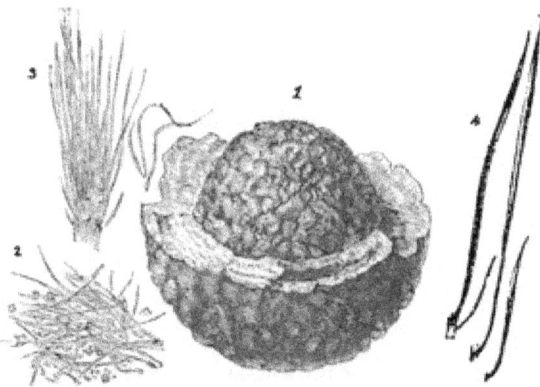

ILLUSTRATIONS OF THE OAT HAIR CALCULUS. COPIED FROM THE INTERESTING WORK
ON CONCRETIONS, BY PROFESSOR MORTON.

1. A section of an Oat Hair Calculus. 2. Magnified Hairs, mixed with crystals of the phos-
phates. 3. Hairs, further magnified. 4. Hairs, so enlarged, as to display their bulbous
insertions and curved forms.

grain in the possibility of such accumulations, the true nature
of which was first pointed out by Professor Morton, and, by

that learned gentleman, these concretions were appropriately designated " *Oat Hair Calculi*."

It has long been known that digestion is promoted by crushing the corn before placing it in the manger. This custom, as a part of the proper process, cannot be too highly commended. But careless horse-owners sometimes purchase the stable provender in a crushed state, or send to have this process performed elsewhere than on their own premises. Such habits are strongly objected to : the horse is surrounded by so much dishonesty, that a prudent man is not justified in trusting the animal's food to the possibility of exchange or of adulteration.

To convey to the reader a definite notion of the very different characters impressed upon various samples of oats, the following illustrations of a few of those which were kindly supplied by a wholesale firm, transacting business at the Corn Exchange, are presented on the following page.

A sound oat should be dry and hard : it should almost chip asunder, and not be torn or broken into pieces by compression. In the autumn months great care is needed to procure sound corn ; the non-professional purchaser is,. perhaps, best protected when he deals for such an article with responsible tradespeople, who, in their business, have a character to sacrifice.

It is a custom to expel the moisture from new grain by drying it in a kiln. It is thereby, in some degree, improved ; but it cannot be said to be rendered as wholesome as sound corn hardened by the natural process. Moreover, oats, badly harvested or damaged by wet, are frequently placed in the kiln, where they are exposed to the sulphur, in order to change or amend their colour. The husks, however, at the conclusion of the process, are seldom all of one tint. If closely examined, indications of the original defect may be discovered on some grains, while others will be of an unnatural whiteness. Kiln-dried oats sometimes betray a shrivelled aspect on that part which is near to the beard; such puckering being occasioned

OATS. KERNELS. OATS. KERNELS.

IRISH OATS, BLEACHED. SCOTCH OATS, SECOND QUALITY.

OATS. KERNELS. OATS. KERNELS.

KILN-DRIED DANISH OATS. FINLAND BLACK OATS.

OATS. KERNELS.

ENGLISH FEED.

by the sudden expulsion of much dampness from the interior. The best test, however, is the rapid rubbing of the sample

12

between the palms of the heated hands; when, should sulphur
have been employed, its peculiar odour will be developed.

The author has been thus careful in describing the signs
which declare the presence of sulphur, because that mineral,
although much employed by ordinary farriers, can occasion the
most terrible belly-ache, gripes, fret, or spasms. This affection
is one of the most fearful to which the horse is subject, and is
the more to be dreaded, as it too often leads to other compli-
cations. Perhaps a greater number of animals annually perish
through causes resulting from spasms, than die under any
other equine ailment.

Besides oats, beans are used as food for the horse; but there
is much dispute as to the quantity a horse can with advantage
consume. Beans should not be given as food until they have
been crushed, and even then they are better if subjected to
the action of steam, which will soften without reducing them
to a watery mass.

Horse beans, as grown in England, however, are very
coarse and astringent substances. No wonder if the large
employment of such produce is found to act upon the bowels;
surprise should be expressed if so harsh a food could be consumed without inducing constipation. The Egyptian bean,
nevertheless, is free from such objectionable properties,
being mild and sweet. A larger quantity of this crushed
and moistened seed might be with benefit presented to the
animal. As at present imported, however, it is very imperfectly harvested. Most samples exhibit the shrivelled and
the discoloured skin, which denotes the sickle was resorted to
before the plant was matured; an error, perfectly inexcusable
in a climate which is for nine months of the year free from
rain.

Might not some sound Egyptian beans be procured? From
these, could not a milder and better species of bean be raised
in this country? The field pea is open to the same condemna-

tion; but field peas are not generally employed in stables. Those used for horses are small and white, of foreign growth, and quite unobjectionable. Tares are given only to farm

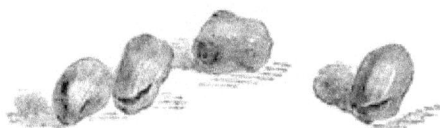

ENGLISH BEANS—A GOOD SAMPLE.

ENGLISH BEANS—A BAD SAMPLE.

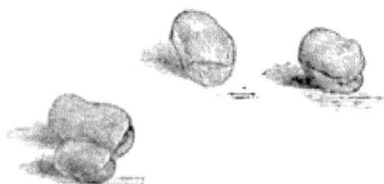

EGYPTIAN BEANS—A GOOD SAMPLE.

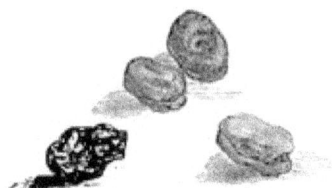

EGYPTIAN BEANS—A BAD SAMPLE

teams; but if this plant possesses only a tithe part of those virtues for which it is accredited, its employment might be advantageously extended. Why should hay be made only of

12 *

grass, which, though admirable sustenance for the bovine tribe, evidently is not equally suited to the equine species? The dropsy of the abdomen and legs it induces in nags, together with the foulness of coat which it engenders, are, perhaps, the best evidence of the injury that attends the long employment of green grass, or even of hay, as a solitary sustenance.

Might not beans, peas, and other leguminous substances, be sown broadcast, and mown when in flower? Hay thus produced would be of all value in the stable, because grass, like corn, whether exhibited green or dry, simply induces fat; whereas leguminous plants all favour the development of muscular fibre, or support the strength of the body. Such hay might be charged a little higher; but then its feeding value and its worth as a promoter of condition would far more than recompense any extra money at which it might be charged.

It may be asked why, if hay produces fat, are the horses of the poor so lamentably lean, since such quadrupeds receive little else than hay to sustain them? The reasons are numerous. The hay such horses obtain is not often of a good quality; and it is to be feared the stuff is not, frequently, presented in sufficient quantity to promote obesity. Besides, this substance leaves the muscular power unrefreshed. The frame being exhausted by a life passed in exertion, the body's weakness effectually counteracts all tendency to fatten.

Beans are not known to be much exposed to deterioration; but oats are liable to an affection of the epidermis, or of the skin, which causes them to be covered with little granules of a dark colour, which the microscope discovers to be fungoid growths resembling a species of very minute toadstools. Corn, when in this condition, is readily recognized by a very powerful musty smell; and the grain, of course, is not adapted to nourish any animal. Musty provender is supposed to engender worms and other unpleasantnesses; but the reviser is disposed to attribute the production of the parasites to a want of resistance in the system, which may be inherited, or spring from a sickly state of the body, or which may be produced by the consumption of unwholesome diet.

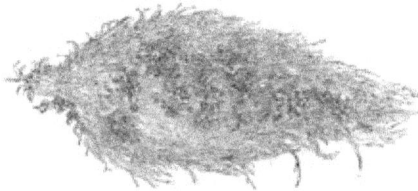

A MAGNIFIED MUSTY OAT.

With regard to quantity in the matter of diet. All animals are not of one size, neither have all horses the same capacity of stomach. It is usual to measure out so much corn as the allowance proper for a horse, and to toss the quantity into the manger, without paying any attention to the desires of the creature. Such a custom may be extremely convenient; but it is very wasteful. Horses differ quite as much as men do in their appetites. By the common practice, one animal receives more than it requires, while another gets less than satisfies its cravings. Some slight notice of the body's necessities should be insisted upon in those who pretend to comprehend the quadruped; and a master should instruct the servant that a creature endowed with life cannot be justly regarded as a manufacturing machine.

Then, as to the times of feeding. The horse is essentially a creature of the night. Man may shake up the straw and lock the stable door; but he does not, therefore, put the quadruped to sleep. Long hours of watchfulness are apt to generate habits of mischief, as well as lead to many indulgences which are no more than the results of want of employment, or the absence of amusement. The solitary confinement, now popu-

lar in prisons, in workhouses, and in some schools, is evidently
wrong in principle; more especially wrong is it when practised upon children; as loneliness, acting upon immaturity of
intelligence, invariably leads to an evil desire, which is, in
penal prisons, spoken of as "breaking out."

To talk of the feelings, the instincts, and the inclinations of
the quadruped, is to earn a character for maudlin affectation.
The populace in the public highways hourly stare at, or care-
lessly pass, spectacles, which, were the general mind really
educated to understand what is before it, should awaken the
keenest commiseration; but which are now viewed as sights
that enliven the prospect. Whence is derived such hardness of heart? Whence springs such general and such a
deep-rooted insensibility? No man seems capable of inter-
posing a voice of expostulation, when the streets display
living and feeling flesh creeping towards its early grave;
when he beholds the animal driven slowly to death; when he
looks upon an animated being, so worn and so dejected, that it
is the last office of humanity to summon the knacker to end a
hopeless misery. The existence of a Society, with two con-
stables, poorly counterbalances a national display of spurs and
of whips. The foremost humanitarian, so the skin be whole,
can afford to gaze upon a lean and spiritless horse, tired
beyond man's most exaggerated conception of fatigue, slowly
creeping before some over-burdened cart, while the driver,
whip in hand, adds his weight to the disproportionate load.
Misery in front,—brutality behind,—and hard-heartedness
around; while a fellow-inhabitant of earth totters onward to
its death!

CHAPTER VI.

THE EVILS WHICH ARE OCCASIONED BY MODERN STABLES.

IT does not require any vast expenditure of thought to discover that life is action; "to be," is synonymous with "to do:" therefore, it is a sheer necessity of existence that an animated being must be doing something. Such is the primary consequence of existence. Thus, to breathe and to move imply one act; since, if the lungs cease to dilate, respiration immediately terminates, and, with it, animation comes to an end. Yet, it remained for mortal perversity to rebuke the first principle of established philosophy, when stables were built, in which a breathing animal was to be treated as it were an inanimate chattel.

Nature, like a kind mother, is to this day endeavouring to teach her wayward children a plain truth, which they may hourly behold enforced by visible examples. The wilful brood appears to be in no hurry to learn. Man still treats the horse as though he honoured the quadruped by enslaving it; and ennobled a life, by conferring upon the animal the title of his servant. He acts as though, by such conduct, sufficient reason were exhibited why he should oblige the creature to resign its instincts and relinquish its desires.

The equine race, when in a wild state, are gregarious, or congregate in herds. Man captures such a quadruped and places it in a stable, built to enforce the extreme of solitary confinement. The plain is the natural abode of the herd; on

their speed depend both their pleasure and their safety.
Man ties the domesticated horse to a manger, and pays a
groom to enforce absolute stagnation upon innate activity.
The "panting steed" is the most timid of living beings.
Man insists the charger is possessed of extraordinary courage;
he declares it delights in the tumult of battle; and he esteems
it a glorious achievement to brutally coerce the timorous sensibility.   The mild-eyed horse is, perhaps, the most simple of
all the breathing beauties which adorn a wondrous world.
Man declares all of the gentle breed have dangerous propen-
sities, and are most inherently vicious.

Before subjugation, the creature fed off the surface of the
earth.   Man builds a house specially designed for the captive,
in which the corn is placed on a level with the chest, and the
hay is stationed as high up as the head.   The animal is gifted
with affections; it longs to gratify their promptings; it yearns
for something upon which its abundant love may gush forth,
—a fellow-prisoner—a goat—a dog—a cat—a fowl;—no
matter what,—so it be some living object, on which may be
lavished that excess of tenderness which, confined to its own
breast, renders being miserable.   Man esteems it his primary
duty to clear the stable of all possible companionship; but the
creature which would rejoice, were it only permitted to worship
its enslaver, he rarely approaches without a loud voice, a
harsh word, or a harsher blow, announcing his presence to
the captive.

The inhabitant of such a prison, a domesticated horse
miserably drags through a shortened life, under human
protection.   The nearest approach it can make to freedom
is its period of labour.   It always rejoices to quit its confinement; but, enfeebled by imprisonment, and subservient
to man's exactions, it ever gladly returns to the place of its
sorrow.   In proportion as its limbs are finely made, and its
actions are graceful, is it prized.   It is never esteemed for its

instincts, or credited with intelligence. It lives in so limited
a space, that, in comparison with the dimensions of its abode,
a man in a sentry-box dwells in a mansion; or a lion in a
cage roams over a domain. A reasonable and an intelligent
being commands his horse should be fastened to such a spot,
and supposes that a living organism is to endure the confinement, which does not permit the body to turn round; that
animated functions are to exist where most ordinary exercises are rendered impossible : nevertheless, he anticipates the
creature will appear bounding with health, in answer to his
requirements.

To be sure the prisoner, although its head be fastened (a
restraint not imposed upon the most savage of carnivorous
beasts), is permitted now to bear upon one leg, and then
to change it for the other. It may perhaps lie down or stand
up, without provoking chastisement. Neither head nor tail
is forbidden a *proper* degree of motion. But at this point all
indulgence is exhausted. It is tied to a rope two yards in
length; but it may not go even to the extent of its tether;
neither may it move close up to the manger; both acts are
equally unpardonable : a *properly* behaved animal should stand
quietly in the centre of its compartment, and always remain
there when not lying down.

It is beaten, if its head be raised just to peep over the
paling, to exchange a rub of the nose and to give, as well
as accept, a warm stream of fragrant breath to and from its
nearest fellow-misery. It must taste the full flavour of its
captivity : no trivial act may distract attention from the
horror of its position. It must lie down where it stands; and
stand where it laid down. It must not display the grace and
ease of motion with which it has been endowed; nor must it
indulge the kindly feelings Providence has gifted it with. If
the owner of the horse does not recognize the exquisite adaptation of sight, so as to infract the minutest particle, and to

view the most distant object, the sensibility of hearing to which movements are audible, when to the duller perceptions of man no sound vibrates on the air; the keenness of scent, which can appreciate qualities in substances which to human sense are devoid of odour; the fleetness of motion which was permitted as a protection, the ease of which the most perfect machinery has failed to rival;—how can he expect his servant to inquire whether such attributes were given by Nature, only to be fastened by the head or to be confined within a space in which absolute stagnation must ultimately induce bodily incapacity.

Such a true "Vis inertiæ" can alone be varied by the hours of labour and the periods of feeding. All pastime is unlawful; the most innocent amusement must be practised silently and in secret. Certain animals, however, try to get through the long hours of enforced idleness by quietly nibbling at the

NIBBLING THE WOOD-WORK.

topmost rail of the manger. Large portions of tough wood are often removed after this fashion; and, to him who can rightly interpret signs, a thick post bitten away, fibre by fibre, will present melancholy evidence of that longing for employment, which could induce so great a waste of perseverance; for animals are naturally great economists of labour.

Other prisoners will endeavour to cheat the time by licking

their mangers, apparently in the hope that some stray grain of corn may have escaped previous attention. The soft tongue of the horse, passed over the hardened surface of the wood, occasions no noise. Often a few grains will have lodged in the corners; then the effort to displace these affords a long game. Others, from want of something to do, or from finding impure air and inactivity do not, in accordance with the general doctrine, promote equine digestion, learn " to crib;" a few, from the operation of the like causes, become perfect as " wind suckers." All "speed the weary hours" as they best can; and many heads are turned round to discover if it be feeding-time again; not that they are hungry, but eating is an occupation, and they sadly wish for some employment.

Certain quadrupeds, under these circumstances, adopt a habit, which is the more remarkable, because hours of tedium have generated the like indulgence in human beings. Mortals, when compelled to remain stationary, and forced to preserve silence, often strive to kill time by rocking to and fro, or by " see-sawing " their bodies. Such a pitiable excuse for amusement is very common among the little people whose undeveloped limbs are perched on high forms, and in whose hands are fixed very uninteresting primers, from which the infant mind wanders into vacuity during the hours of imprisonment which occur in those pleasing places termed " Preparatory Schools." The horse, also, when forbidden the pleasures in which Nature formed it to delight, will move its head methodically from one side of its stall to the other, and will continue thus engaged for hours together.

So exciting a pastime, most sane people might deem to be harmless enough; but by the arbitrary notions of rectitude entertained within the stable such custom is punished as a vice. A horse which "see-saws" is said to weave, and weaving is, by grooms, esteemed highly culpable, and is usually corrected with the lash.

Can human perversity conceive a life without a pastime, and vexatiously impose this terrible fate upon the creature whose existence is devoted to man's service? When in the field, the horse is never idle. The only amusement of the simple animal ies in its perpetual occupation. What a despairing sorrow must therefore afflict such an existence, when dragging through its time under the fostering care of the enslaver. Yet how proudly do some intellectual beings boast of their stables and of the ceaseless attention lavished on their studs. What is it this assiduity realizes to the creature on which it is expended? Stagnation to the active, and solitude to the gregarious. Movement draws down punishment, as it were a fault. Any attempt to while away the tedious hours is esteemed "a vice;" sensation must be checked, and feeling, man insists shall be suppressed. But who, among the millions of intellectual masters, sufficiently understands the quadruped, over which they all usurp authority, to regard the huge bulk of that endurance, as the embodiment of the acutest form of every possible earthly misery?

Perpetual inaction also occasions waste of food: the horse, wanting exercise, stares at his provender, but has no appetite; the hay piled up before it is nothing more than matter out of place.

Desire is needed to give value to such abundance; and a non-reasoning being cannot be expected to prize that which it does not require. It cannot eat; but it lacks amusement. The hay is before it. In sheer idleness, a few stalks are pulled from the rack. Of these, one may be leisurely masticated; but the remainder, after having been twisted about by the lips, are allowed to fall upon the litter. The sport is followed up until the rack is emptied; and the creature is a little happier, under a conviction that it has escaped from absolute stagnation.

The sin, if there be any, certainly must remain with the

man who piled up the provender before the animal which was
without an appetite.

Simple natures, when entirely disengaged, generally make
their own employment, and that employment, being intended
for a passing amusement, commonly consists of what thrifty
people designate "mischief." The knowledge, that dis-
pleasure will follow upon discovery, may spice the proceeding
which otherwise might want interest. At all events, so it is
with children; and it may be thus with animals. When a
heaped manger is before a satiated quadruped, the impossibility of feeding makes the creature meditate upon the uses to
which the grain can possibly be applied. None can be discovered. The head of the captive is tied, and the manger is
fixed. At length, in carelessness of spirit, a mouthful is
taken from the heap. The portion cannot be swallowed, so
the lips are moved, and, as they part, the corn falls over them

WASTING CORN.

upon the ground. This may not be a very exciting recreation;
but the prisoner is restless with repletion. It cannot sleep;
and the grain passing over the lips, in which equine feeling
concentrates, produces a slight and a novel sensation.

Can any man seriously pronounce that an animal, standing

in enforced solitude and compulsory idleness, is to blame for
such conduct? Boys, during their school days, when wanting
appetite, or having unnecessary food before them, will not
they, in satiety, play with needless abundance? Are men to
demand that prudence from an animal, which we should certainly not anticipate in the young of our own species? Yet
the child enjoys a certain amount of confidence; and its mis-
doing is, therefore, aggravated by a certain abuse of trust.
The horse is confined between boards, and enjoys not the
smallest personal liberty. The severity of captivity argues,
that no reliance reposes upon the captive's discretion. All
responsibility is lost when all freedom of action is denied.
Yet the poor prisoner is cruelly beaten for playing with food,
although the true fault rested upon him who was too idle to
give the exercise which would have generated appetite; and
was too lazy to proportion the animal's sustenance to the
requirements of its situation.

Another so-named "vice" of the horse is frequently the
occasion of more serious results than any of the before-men-
tioned accidents. No person has hitherto explained why the
skin should be more irritable by night than during the day-
time. Such, however, is the case with horses, as it is with
men. A quadruped, in the morning, is often found disfigured
by the hair being removed from comparatively large surfaces.
Itchiness has provoked the animal to rub itself against any
prominence, or to scratch its body with the toe of its iron
shoe; this indulgence has caused the blemish.

Itching and scratching are numbered among the worst
"vices" of the stable. Such faults, however, are only discovered in their effects; the groom never estimates, when
flogging an animal for this wickedness, how far the abhorred
sin may have been produced by stimulating diet, by want of
exercise, and by impure atmosphere. No! he clothes up the
body of the animal; shuts every window; stops every cranny;

and locks the stable door for the night. The last meal being consumed, and the quadrupeds not being inclined for sleep, they one and all begin to itch. Legs are nibbled; necks are rubbed; and tails are lashed. At length one is sensible of an irritation behind the ear. The head is turned toward the side; the body is curved to the full extent; and the hind leg brought forward. Then, the groom not being present, the toe of the hind shoe can touch the part, and the horse luxuriates in a hearty titillation.

When the head was turned towards the quarters, however, the collar-rope, being attached to the halter, was also stretched in that direction. The hind foot having performed its office, a desire is felt to return it to the natural position. The attempt is made; but this is found to be impracticable. The creature strains against the opposing force, but its struggles only render its comfortless attitude the more fixed. The truth is, that while devoted to the act which allays cuticular irritability, the pastern has slipped over the collar-rope. Such a mishap not only fixes the leg, but fastens the head. With the neck bent and one leg disabled, the animal cannot exert half its power; neither can simplicity comprehend the source of its unnatural constraint. Long continuance of the position becomes painful; alarm seizes upon timidity; the struggles grow desperate; and the poor quadruped, at length, is cast with terrible violence upon the straw which had been shaken down for its repose.

The animal is lucky which should be overthrown in a limited space and escape serious misfortune. It can hardly encounter such an accident and rise from the ground uninjured. The slightest consequences are contused wounds or fractures of small osseous prominences. The worst result, however, usually follows the body being forcibly contorted throughout an entire night. Bones have been dislocated, or a limb has been so sprained as never to have recovered its functions.

Necks have never afterwards been restored to their pristine
grace of motion ; and, in short, a valuable servant has, by
such a misfortune, been so " wrenched from its propriety " as to
be rendered utterly useless.   Nevertheless, the groom will
persevere in hailing the fate of an animal which has been cast
in the collar-rope as a just punishment induced by the sufferer's inveterate " vice."

CAST IN THE COLLAR-ROPE.

Carters are open to complaint, because their horses are
" cast in the halter " even to a greater degree than those of
town grooms.   In agricultural districts, it is a common practice to turn the teams out to graze during the night, and to
take them from the field to work in the morning.   Some
animals, however, prove troublesome to catch, preferring the
cool grass and partial liberty to exhausting toil upon an arid
roadway.   To facilitate the capture of such quadrupeds, many
carters, when freeing the creature, will not remove the halter,
but suffer it to remain, because this affords a ready hold for
the person who fetches in the horses on the following day.
The result is easily anticipated.   The ear itches.   The foot,

scratching the part, gets entangled, and that which was a valuable horse on the previous night, is found, in the dawning light, to be a disabled cripple, or a worthless carcass.

The sane mind will, however, behold in this misfortune only a startling proof of the folly which ties the head to a manger, and leaves the animal to the hazard of a fearful accident.

It may be urged that the size of the horse's body necessarily limits the dimensions of its abiding-place. This is a strange reason; but it is one commonly used among architects. We, however, do not apply the principle to our own race. Because the Horse Guards are tall men, we do not insist they should sleep in infants' *cots*, or wear the clothes of children. Giants are not forced to inhabit the houses fit only for dwarfs. Neither do we carry out the maxim with other creatures. Large rabbits, boys put into large hutches. Were smaller horses desired, ponies, even no higher than full-sized dogs, are not scarce. But greater weight and strength enable the quadrupeds to perform larger services. Does it not seem like meanness to select size for our own purposes, yet, where the creature is concerned, to make size a motive for stinting the necessities? The horse is useful to man in proportion to its magnitude; and the poor slave, therefore, ought not to feel the bulk to be its misfortune!

Other evils remain to be narrated, which are consequent upon the use of the halter rope. The horse may, when in the act of rolling, biting at flies, or other substances such as dried perspiration, which have caused irritability of skin, get his fore leg or hind leg entangled in the halter rope; if the latter occurs, he becomes what is called cast, and the repeated efforts to rise usually draw the rope more tightly around the hind leg, which occasions a severe wound. Should nothing but a bruise or abrasion be present, the groom undertakes the immediate relief.

13

Having disengaged the rope from the injured limb by untying the knot below the sinker, the head is released from the manger.   But it is not usual for quadrupeds, after such a misfortune, to rise immediately; it is a good plan, conse-

A HORSE CAST BY ROLLING IN THE MANGER.   A HORSE BEING RELEASED FROM THE PREVIOUS SITUATION.

quently, for the groom to obtain an assistant, and, with this combined aid, to drag the body and hind legs away from the stall posts and stall partitions, and thus place him in a position to rise without difficulty.

The remedies applied to all injuries (excepting fractures) which occur in the stable are equally simple, and few in number.   These consist of a lotion, composed of two ounces of tincture of arnica, which is put into a pint bottle, to be subsequently filled up with water.   This is used till all symptoms of bruise or swelling have disappeared, after which another lotion is to replace the first.   This last is formed by adding one grain of chloride of zinc to every ounce of water, or one scruple to each measured pint of fluid.   These lotions are to

be applied frequently, not directly, to the injury itself, but a
sponge, saturated with each liquid, is to be squeezed dry
above the sore, the moisture being allowed to trickle over
the wound.

The strongest testimony, however, against stables, as such
buildings are at present erected, is perhaps borne by the
animals which inhabit those places. The horse is a delicate
test, which man would do well to attentively observe when he
is desirous of ascertaining the healthfulness of any locality.
Naturally it is all animation and gaiety of spirit. But, however much these qualities may be esteemed, such equine re-
commendations will soon fade before the joint influence of
impure air and close confinement, although you may groom
and feed at discretion. The natural period of life is diminished
one half, while much more than half of the remaining years is
rendered useless by age, prematurely brought on by inappro-
priate treatment.

A FORE LEG OVER THE COLLAR-ROPE.

CHAPTER VII.

FAULTS INSEPARABLE FROM STABLES.

NO gentleman regards his country seat as finished until
to it ample stabling is appended. A small space,
judiciously employed, can be made to house a great number
of horses.

The architect, being informed of the wishes of his employer,
unhesitatingly asserts that 5 feet, or 5 feet 6, or, in extreme
cases, 6 feet, are considered ample width for stalls. The proprietor agrees to grant the last-named space for the abiding-
place by day of a living horse, and the spot on which rest must
be enjoyed during night by the same huge quadruped.

This decided, the gentleman rubs his hands, and, warmed
by the contemplation of his own liberality, applauds "the
nice arrangements," which he has sanctioned "regardless
of expense." But the carriage-house, he is positive, shall
be built quite large enough. He cannot forget that those
rascals grazed his last new vehicle on the very day it came
home from Long Acre. The accident happened while putting
it into a narrow building. No! Let what will be cramped,
the carriage-house must be spacious.

Thus men take much care of that species of property which,
being damaged, can be repaired for money; but they treat with
neglect, and thrust into unwholesome corners, that life which,
when injured, not all the wealth accumulated upon this globe
could restore to soundness. With the inanimate there is

nothing to remove the full force of blame, which man must accept as his fault alone. The deterioration of such articles, when it occurs, cannot be laid to the charge of any other living being. This renders man more careful of such things. With life, there is always something which can be made to take the weight of culpability from the master's shoulders. The horse was obstinate; it had a bad temper; it possessed a vile mouth; it bolted; it refused; it shied; it reared; it jibbed; it kicked, or, in some way, it resolved not to do its duty. The dumb creature can make no answer to the accusation; and human nature is readily convinced of its impartiality when its errors have been meanly cast upon another life.

The builder is, of course, governed by the architect; the architect is anxious to exhibit plans which shall elicit the approval of the proprietor. So, in the end, those arrangements, upon which the well-being and the health of many lives must depend, rest upon the caprice of an elderly gentleman, who now, for the first time in his life, may give serious thought to such a subject. However, this is the rule, whether a house is intended for a family residence, or is erected as a speculation. The stables almost invariably occupy the space which is left after every other want is satisfied, consequently economy of space is carried into all the detail relating to the stable construction. The doors need not be wider than 4 or 5 feet; and certainly a doorway 5 feet wide affords somewhat ample space for a horse's body to pass through, i.e. if he be quietly disposed; but even so vast a limit will not allow a groom to dispense with every care. An animal may, reasonably, be delighted when it sniffs the fresh air, and it may be permitted to perform a few pranks, as it quits positive stagnation, to make the nearest approach to freedom which its enslaved condition can sanction. Schoolboys do not observe any severity of order when they cast aside their tasks to throng into the playground; yet the youths are confined to

study only for a comparatively short period. But what must
be the feelings of the steed when leaving the heated stable
and narrow stall, where it has probably been imprisoned for
twenty-two consecutive hours?

Who among us, if he had the power, would check the
graceful prancings and elegant curvettings by which a simple
nature announces its sense of happiness? To human feeling,
an idea of having to carry another's weight, in the direction
and at the pace the rider pleases to command; to have a sharp
bit pulled against the tender angles of the mouth, conjures up

"DOWN IN THE HIP;" OR A HORSE WITH THE BONE OF ONE HIP FRACTURED.

an image calculated to awaken no special delight. But long
imprisonment may induce that eagerness to breathe the air of
heaven, which may possibly render the prospect of labour,
beyond the confines of its jail, welcome to the captive.

Quadrupeds have been injured while passing through the
widest of modern entrances. The pleasure of escaping from

the tedium and from the faintness of actual stagnation gene-
rates a joy which banishes the sense of prudence.   All feeling
and every caution appear to be engulfed in the exultation of
the moment.   The horse dances as it walks; the tail is gaily
whisked; the neck is arched; the mane is shaken and the
body is twisted by those numberless undulations which have
often excited the admiration of enthusiastic spectators.   If,
during one of these expressive movements, the trunk should
be inflected more than the seven inches which the five feet
allows, or the animal, influenced by the impetuosity of excitement, should come in contact with the door-post, the consequence may be fearful.   The possibility of check, certainly,
does not enter the thought of the joyous creature.   The blow
is proportioned to the heedlessness which induced it.   A bone
can be fractured on such an occasion; nor is it an unusual
accident.   Most horses, which are beheld with one " hip
down," have had the deformity produced by striking against
the post of the stable-door.

"Down in the hip" is a groom's phrase, and merely signifies that one of the prominences of the haunch-bone, or,
employing anatomical language, that one of the inferior
spinous processes of the ileum has been broken off.   This
osseous projection is of great importance to the value of the
quadruped; it gives origin to numerous muscles, but more
particularly to the powerful extensors of the hind limb.
That ease, grace, and rapidity with which the member should
be moved are by this misfortune destroyed, and the animal
is thereby unfitted for the more highly-esteemed half of its
future services.   By the accident it loses caste, and moves
downward in the scale of equine employments.

This terrible affliction to the life principally concerned, may
also be occasioned in another manner.   Grooms, when leading
a horse from the stable, commonly walk by the side of the
animal.   Such persons are usually fully dressed to attend

their masters, when called upon to perform this duty. Should the creature in its joy, when passing through the doorway, touch the coat of the domestic, such familiarity elicits the utmost indignation. A loud word, a kick, or a blow instantly resents the insult. The animal, in terror, skips about to avoid further punishment. The door-post is struck; the haunch is fractured, or the pain is inflicted which renders the creature, with its retentive memory, ever after fearful when passing through an entrance.

The ordinary life of a domesticated horse is so monotonous, that recollection of events cannot otherwise than be retained. The animal, subsequent to such a calamity, even though no bone should be fractured, cannot gaze upon a door with calmness. In future, alarm is exhibited whenever an entrance has to be approached. It cannot enter or quit its abiding-place without displaying those symptoms of terror which, to the groom, are the representatives only of inveterate " vice." The most violent or the blandest of tones cannot restore placidity to the brain, which is troubled by fearful recollections. It is useless to coax, to threaten, or to punish : the animal has no ability to assume its former quietude, when passing through the terrible opening. But it strives to brace up its nerves for the performance of the necessary act. All its resolution is summoned till, maddened by excitement, it wildly dashes through the entrance, dragging after it the boy to whose custody the more dangerous quadrupeds are, usually, entrusted by the prudent sagacity of stable-men.

With regard to the subject, on which recent comments have been based, horsemen should order their servants never to walk through a doorway by the side of the quadruped, which general belief supposes to be led through such openings.

A boy should not be employed in such an office. Prior to leaving the building, the groom should place himself directly in front of his charge. A short hold of either rein should then be taken in each hand. When there located, he can with ease and certainty guide the head of the horse. The motions of the head regulate the movements of the body, and having the controlling power entirely at his command, the servant should commence to back slowly out of the stable.

LEADING A HORSE THROUGH THE STABLE DOORWAY.

Supposing this obvious recommendation to be adopted, should any symptom of alarm or any disposition to display restiveness chance to be exhibited, progress must be immediately stopped; nor ought it to be again resumed, until the animal has thoroughly recovered its composure. No matter how long a period may be required to restore tranquillity, the groom should, contentedly, continue stationary till every sign of timidity is banished or dispelled. In this manner man and horse should leave the building, nor ought the groom to quit his post until the doorway has been cleared.

If solitude and confinement generate disease in man, why should not similar causes produce similar results upon the horse? They do, and evils, too, of a local nature, result from causes which, if properly understood, could be easily removed. The fore legs of the stabled horse are always the first to yield, and sometimes the hinder limbs are involved in disease. Should one of these last be injured, the animal evidently pines in torture, and wastes in condition; but with the fore legs affected the horse seldom loses flesh, and in many instances he has been known to grow fat. Physiologists account for this peculiarity in various ways. Some appeal to the greater proximity of the anterior extremities to the heart, or to the centre of circulation. That, perhaps, is the generally received doctrine; but as the free circulation of the blood is essential to the healthy functions of the nerves, it is difficult to comprehend why nearness to the heart should deprive a nerve of its ability to communicate sensation. The head is supposed to be rendered conspicuously sensitive, because of the great proportionate quantity of blood which circulates in that region. The pretended rule, therefore, will not bear the test of general application; it must be discarded as an assertion boldly put forward to cover ignorance.

The fore feet of the horse are those portions of the frame which have to endure the utmost limits of mortal perversity. The flooring of the stall invariably inclines from the manger to the gangway. The hind hoofs may, should the animal hang back the full length of its collar-rope, rest in the open drain with the toes downward; or the hind hoofs may, in some cases, stand upon the gangway, the width of which the guttor defines. The front limbs, however, can scarcely change their position. The hoofs must rest upon the slanting bricks, which incline the anterior of the foot in the upward direction. The fore-legs must sustain, and continue subject to the unnatural stress of their enforced position. This silly and arbitrary arrangement in some measure accounts for the fact that the front limbs of the horse are the first parts of the body to fail, for these parts never, in the stable, are capable of rest, nor can they be sensible to ease.

It has, of late years, become the general practice to bleed the horse from the sole of the fore foot. When such a custom is adopted, the first portion of blood extracted is commonly cold as spring-water, or from thirty to forty degrees below the standard recognized as " blood-heat." Now a certain warmth is imperative to the existence of vitality, which is arrested so soon as the natural heat of the body is sensibly diminished. The functions are stayed when any region has parted with its caloric. Dentists take advantage of this fact when, after having employed the chilling process, they extract a tooth without pain. Cold, therefore, which can destroy sensation in the human jaw, likewise renders the foot of the horse insensible to agony.

But why is the fore leg subject to a degree of cold which does not also affect the hind extremity of the animal? Because the stable permits the hind limbs to enjoy the greater freedom of action. These may be in perpetual motion; for the posterior members are situated at the boundary of a circle, of which the ring of the manger represents the centre or fixed point. Has the reader ever beheld a column of soldiers move in obedience to the officer's command to "Wheel"? The man at one end of the line can hardly run quickly enough, while he who is placed at the opposite extremity is troubled to be sufficiently slow in his movements. Now the hind legs of the horse represent the man who has to scamper, and are

sensibly exerted whenever the quadruped "comes over": the
anterior extremities are types of the soldier who scarcely
moves, for very seldom are these members necessitated to
change their position. Their stable office is to uphold the
body, and to remain fixed while the toes are inclined upwards!
Were the motion permitted to both extremities equalized, the
fore limbs would naturally be the warmest, since the great
distance from the heart and the greater angularity of form
must render circulation of caloric within the quarters much
more tardy.

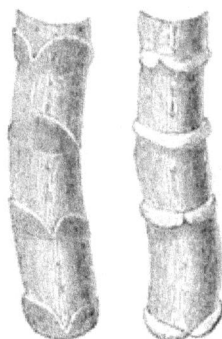

VALVES OF THE VEINS IN THE LEG.

1. The valves of the vein laid against the side of the vessel by the upward current of the
blood.

2. The valves projected from the side of the vessel by the attempted retrogression of the
vital stream.

But why do not other parts become as cold as the fore
limbs, when all belong to the same body, and all derive their
heat from one common medium, or from the same circulation? The veins in the legs have valves. Then, if these
vessels are so provided, and the distribution of warmth is
one of the purposes of the circulation, why do not the valves
favour the return of blood from the foot, and thus generate
heat within the member? When answering the foregoing

inquiries, the reader's patience is entreated, since the reply, to be intelligible, cannot also be concise.

Anatomy affords the best explanation of the peculiarity. On removing the horny case from the hoof of a dead horse, a secretive membrane is exposed; this membrane constantly renews the horn. Beneath the secreting surface a complex mesh-work of large veins is discovered, which, by their size, inform us they serve as receptacles or sinuses quite as much as vessels. These veins have no valves, though such are commonly present in other tubes of the same class. The absence of this provision is in them remarkable, because the blood has to move against gravity : valves are a means instituted to favour the current under circumstances of this nature. Valves are composed of duplicatures of the lining membrane of veins; when the venous current flows towards the heart, these valves, by the impetus of the stream, are forced upward, and remain close against the sides of the vessels ; but, should the slightest retrogression of the current be endangered, the backward motion of the blood carries the numerous valves outward or downward, and effectually locks the interior of the veins.

The anatomy of the foot, however, proves the horse unsuited to confinement. The animal was created to dwell upon the plain. The foot, for its health, requires perpetual motion. When free, or before man subjected it to his convenience, every bite the creature took necessitated a fresh step. The mesh-work of veins was large, the vessels freely communicated with each other, and were devoid of valves, that the blood might readily flow into, while it might as readily be expelled from the tubes ; and because, in the habits of her creature, Nature had established a force which rendered the development of valves unnecessary. The horse, as it progressed, alternately lifted the foot from the earth and rested it upon the ground. When the hoof was raised, the blood rushed into and filled the mesh-work of veins. When the foot was again placed upon the soil, the superimposed weight squeezed the vessels between the bones and the horn, thus pumping out the blood, or forcing it towards the heart.

Blood, which has become cold, has lost the first of its living properties. Blood, deprived of heat, cannot support health, or supply secretion. Hence the feet of stabled horses, notwithstanding the care of science, the numerous applications, and the endless variety of shoes, all of which are designed to benefit, the hoofs generally become diseased. The quadruped of the agriculturist, although it be neglected and badly shod, yet, because of its slow or constant work, and habitual freedom in the field, usually exhibits feet which are sound and open. The donkey, though much abused and shamefully treated, rarely inhabits a stable, and more seldom enters a stall. Its feet become misshapen ; but the curse of the gentleman's steed, foot lameness, and especially navicular disease, are all but unknown among this tribe of the equine race.

The foregoing statement also affords an explanation, why the most valuable or the stabled horse is so frequently afflicted with contracted hoof, with brittle hoof, with an unhealthy secretion of horn, and with the various other ailments which may be classed under the diseases of the foot. It likewise supplies the most clear reason for the beginning of that disorder which has been denominated "the curse of good horse-flesh,"—Navicularthritis, or ulceration of the navicular bone. Bone is slow to take on morbid action, and ulceration is the accompaniment of low vitality. When the circulation is retarded, the animal powers are enfeebled. Ulceration, affecting a lowly organized structure, is that which a pathologist would anticipate as the consequence of prolonged inaction. It is impossible to say what evils the continuance of such a condition may not induce ; but sand-crack, seedy-toe, and various painful affections, can be clearly traced to have thus originated.

The effect of the stable, also, accounts for the farmer riding

his nag for many years, while few gentlemen approve of a
horse for saddle purposes after it has passed the sixth summer,
notwithstanding their animals are better groomed and more

THE VEINS OF THE HORSE'S FOOT.

The mesh-work of veins, without valves, which are situated immediately under the secretive
membrane of the hoof.

carefully fed.   In the country, farmers' quadrupeds are generally turned into the field, and have to walk for their living.
Grass is a poor food; but the constant exercise keeps the
creatures in sounder health than can be maintained by better
sustenance combined with perpetual confinement.

An absolute necessity for the constant movement of the feet
is to be deduced from the arrangement of the vessels.   The
arterial blood falls almost perpendicularly down the fore limb,
while the venous blood has likewise to ascend against gravity.
This arrangement rendered imperative some propelling force
to return the effete fluid; hence the necessity for the perpetual
employment of the squeezing or pumping action of the hoof.
The habits of the animal to graze only from choice portions of
the herbage occasion a vast distance to be traversed; but such
leisurely sauntering was by Nature kindly intended to keep
sound that portion of the frame on the integrity of which the
safety, the welfare, and the pleasure of her creation was
dependent.

"Certainly," the reader may exclaim; "but if the warmth
of the body is dependent upon arterial blood, the coldness of a
part cannot be accounted for by stating the facility afforded

for the oxygenating current reaching that which is chilled."
Very true. But before any substance can fall down, the space
through which it has to pass in its descent must be made
clear. The quickness with which the arterial blood reaches
the foot is, consequently, regulated by the speed with which
the venous current is expelled. The hoof of the stabled horse
is constantly congested, or the effete blood accumulates within
the horn; because motion in the venous stream is impossible.
The current hardly stirs, and the fluid, by stagnation, becomes
cold. Did the possibility of action allow the pumping-force
fair play, then the fore foot would, doubtless, be as warm as
other parts of the animal's system.

Anatomy demonstrates these facts; but the habits of the
quadruped have never been attentively noted. Had the instinctive promptings of its desires been studied with a wish to
profit by such instruction, stables had been erected for some
better purpose than to closely confine an active animal, and to
illustrate the earliest principles of surface-drainage. As it is,
a building has been raised totally inadequate to its pretended
uses, and one in the arrangement of which the convenience of
man has alone been consulted. In such a place a horse has
for ages been imprisoned. It is true the captive did not
thrive; yet this consequence was rather excused than in-
quired into. Humanity has endured loss, disappointment,
and vexation; but pride found it more agreeable to accuse
the works of heaven with the results of its own culpability
than to suspect the adequacy of its own institutions. Nature
has in vain laboured to instruct the waywardness of conceit.
Mankind could endure all evils before it could afford to
question the perfectibility of mortal invention.

Horses, when disposed to remain stationary, always select
ground where the fore feet can occupy a position lower than
the hind legs. In stables this inclination is reversed, the
hinder limbs invariably resting on an inferior level to what

tho fore feet range upon. The motive upon which the
dictates of nature are outraged is the facility which a floor
slanting in the backward direction affords for surface moisture
to flow into the open gutter that runs along the extreme

HORSES, WHEN FREE TO CHOOSE, ALWAYS STAND WITH THE FORE HOOFS ON A LOWER LEVEL THAN
THAT OCCUPIED BY THE HIND FEET.

margin of the gangway. Science, evidently, has not been
consulted in an arrangement which sacrifices the health and
the comfort of an inhabitant of the stall to obtain so obvious,
gross, and poor an advantage.

A creature standing on a slanting floor, with the head
pointing to the most elevated part of the incline, occupies the
same relative situation which the body would possess were the
quadruped journeying up the side of a hill. By the sloping
nature of the ground, the weight of the frame is partially

14

removed from the insensitive bones; and to such an extent as the osseous structures are relieved, is the burden thrown upon the flexor tendons, or upon the back sinews. It is imperative for the health of bone that it should endure almost continuous pressure. On the other hand, tendon or sinew feels no pain from occasional tension; but pressure, if long sustained, produces the acutest agony. When one structure is denied to fulfil the uses for which it was created, and another structure is condemned to discharge services for which it never was

WHEN CONFINED TO THE STALL HORSES GENERALLY STAND WITH THE HIND FEET UPON A LOWER LEVEL THAN THE FORE HOOFS OCCUPY.

designed, the first soon degenerates from not having sufficient employment, while the second speedily becomes disorganized from the necessity to perform too much labour.

Bone, tendon, and cellular tissue almost compose the shin and the foot of the animal. Horsemen know how difficult it is to make and keep the legs of a stabled quadruped hard and fine. It is, however, folly to rub and to bandage while inactivity is permitted to generate congestion. No application can possibly destroy the effect while that cause is allowed to be in operation. Nor can the foot secrete sound horn while the exercise, which is imperative for health, is withheld. No shoe can give that which is dependent upon motion. There are many more pieces of iron curved, hollowed, raised, and indented than the author has cared to enumerate. All, however, have failed to restore health to the hoof. Some, by enforcing a change of position, may for a time appear to mitigate the evil; but none can, in the long run, cure the disorder under which the horn evidently suffers.

Anointing the hoof, or using various stoppings, is equally fruitless. Both leg and foot, after a day of hard labour, only return to the stable to undergo more excessive, because more continuous, fatigue. The sloping pavement renders ease an

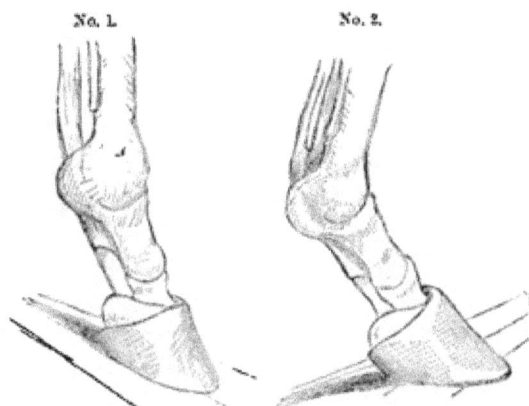

THE STRESS ENDURED BY THE DIFFERENT STRUCTURES WHICH COMPOSE THE LEG OF THE HORSE
IS DEPENDENT UPON THE POSITION OF THE FOOT.

impossibility. The exhaustion cannot be banished from limbs forced to occupy such ground. Longer rest but induces additional enervation.

The reader will observe that, in the sketch No. 1 the bones rest one upon the other. That arrangement ensues when the animal descends an incline. There can exist no man but must have enjoyed the ease which is imparted by walking down a

14 *

slope. Every person must also be acquainted with the fatigue consequent upon ascending an acclivity. The effect is generally explained by stating that, in one case, progression is favoured by, while, in the other, it is made in opposition to, gravitation. Such a cause, certainly, is in operation; but the different structures on which the strain reposes, when moving in opposite directions, to the author's mind supply a better illustration of the fact.

Do not muscles, and does not tendon participate in the burden which is upheld by bone? Assuredly they do; but in various degrees. No limb can move unless some muscle contracts. Every muscle in the body arises from bone, and is inserted into bone by the interposition of ligamentous fibre. Before a member can be elevated or depressed, some muscle must drag from some bone, that it may move some other bone more distantly situated. Then, tendon cannot escape strain, since in no possible attitude is every portion of the frame in absolute rest. Motor muscles, however, generally exist in pairs. They are spoken of as elevators and as depressors, or, as flexors and as extensors. Their uses are opposite, but not opposed. When one set works, the other is inactive.

The bones in the sketch, indicated by No. 2, evidently press against the backward tendons. Such a position, if long maintained, leads to fatigue, and ultimately induces pain. Man cannot enjoy rest under such a condition of parts; though both tendon and muscle are benefitted by brief tension, continuous strain soon exhausts either structure. The reader must have beheld two travellers meet upon a mountain's side. One shall be descending from the heights, the other is ascending from the valley. But while the men converse, they do not hold their relative positions one to the other. Each, without thought or reflection, exchanges it for the horizontal situation; while their dialogue lasts, both present their sides to the spectator.

This is precisely what many horses learn to do.   Much
indignation is always excited in the groom's bosom because an
animal, prompted by its instinct, has discovered a method of
easing its limbs and of saving the master's property from
injury.   Standing for hours upon an acclivity, however gradual,
throws stress upon the back sinews, and must pain the tired
limbs.   To counteract that effect, the animal turns the head

STANDING ACROSS THE STALL.

from the manger, and stands across the flooring of the stall,
after the same plan as actuated the two travellers when they
paused upon the mountain-side.   But the conduct which, in
man, draws forth no remark, when exhibited by the horse is
abominated by a virtuous groom as the declaration of inveterate
" vice."

Pitiable vice! It is melancholy to behold a man cruelly punish an animal for a reasonable act. But heavy castigation does every horse receive that is guilty of exercising the instinct with which Heaven has endowed it. The groom, being excited to resentment, grasps a stick, and deals well-aimed blows, while his voice shouts forth harsh words, which pain and terrify the patient creature, whose only faults were too much sense and too great feeling.

When a horse is terrified, danger is likely to ensue in exact proportion to the smallness of that space which can be commanded for the display of its alarm. The timidity being excessive, of course, the contortions of the body are equally demonstrative. The animal dashes about, regardless of its own safety, and heedless of those around it. It sees nothing; it can remember nothing, save only that some horrid torture is imminent. Its struggles are wild efforts to escape. In the momentary panic, it may break, or it may damage anything. It may kill any person who shall stand in its way, or, in the furore of its agony, it may, through misadventure, do serious mischief to its own body.

Such consequences are always to be expected when a horse is beaten within the stable, while the head is fastened to the manger. By the latter circumstance the probability of an injury is increased. Harm, however, to his employer's property, danger to his own person, and peril to the safety of his charge, the groom despises, or willingly hazards, rather than allow an odious "vice" to escape correction! No severity, however, can teach a quadruped not to seek the ease which it has discovered the means of realizing. When the groom is absent, or during the night, the act of "wickedness" is always renewed, although, in the presence of its attendant, the indulgence may be suppressed.

Slanting pavements, likewise, instruct horses in the practice of other habits, which the groom, in his peculiar sphere of

mental elevation, cannot otherwise than recognize as "vices."
As such, he punishes their exhibition without mercy. Some
public-house companion may visit the stable-man, while he is
dozing through the afternoon upon the locker. Most servants
notoriously have no choice between stubborn duty and the
relaxation of "pipe and pot." The groom is always the
ready victim of temptation, and, upon the slightest persuasion,

THE HIND FEET ARE EASED IN THE GUTTER.

quits the stable for the parlour "over the road." Some sad
and patient animal may have been silently watching, longing
for the man's absence, during a considerable period; no sooner
does the creature hear the door slam, than it begins to take
small steps backward. The horse thus feels its way, till the
sudden fall in the pavement announces that the posterior hoofs

have reached the gutter, within the hollow of which the toes are immediately depressed.

Such an attitude being attained, all stress upon the flexor tendons is removed from the backward legs. The bones, while the toes can be depressed, sustain the weight of the haunches. Partial ease is thereby secured, and with the new sensation a numbing torpor creeps over the animal. Its feelings are soothed by present pleasure, and the senses, thrown off their guard, grow dead to all outward impressions. The victim of former ages, when taken from the rack, must still have endured agony; but the lull occasioned by the cessation of acute torture threw the sufferer into a lethargy, which is reported to have resembled the luxury of sleep. So is it with the horse. The fore feet are still undergoing torment; but, under partial relief, the animal seems to doze, or becomes unconscious to the facts around it.

The horse is tranquilly luxuriating, and cozily revelling in the moments of forbidden ease, when the groom quietly returns to the stable. His eyes rest upon the animal who dared to indulge in a little ease; the position infuriates him, and with harsh words and whip he commences to lash the culprit, for such he considers him to be.

The stabled horse, however, has not only to stand upon a slanting pavement through the day: it must throughout the night lie upon a similar incline, rendered slippery by a covering of dry and polished straw. Did the reader ever attempt to repose upon a bed slightly out of the horizontal? The body cannot rest on such a couch. The sensation communicated is, an incessant fear of slipping off. The sleeper is constantly wakened up, with a vivid impression that he is falling, or has fallen, on to the floor. The night is passed in discomfort. But, what is the excitability of a human being, when compared with the excessive fear which haunts the most timid of all created lives.

Man, when in a bed of the above description, naturally grows restless; the bedclothes are disturbed, and the body laid in an opposite direction. All will not allay anxiety; at last, the would-be sleeper is obliged to remain contented with occasionally nudging himself higher on to the pillow. Like man is the horse in many things, even as though the animal studied and mimicked its master. Yet, the inflation of pride hails the resemblance as an insult, and regards animals as things created for use, and doomed to be subservient to the caprice of mortal pleasure.

Precisely as man would behave, did he chance to get upon a slanting bed, the animal conducts itself, only with such difference as the circumstances enforce. The human being reclines his head upon a pillow; but the horse sinks the head while it slumbers. Man, therefore, nearly touches the board, situated at the topmost part of his resting-place. Three feet, or even a larger space, may divide the quadruped from the stable wall, which forms the extremity of its couch. The floor, on which the creature lies, is strewn with straw. That condition, however, rather aggravates the inclination of the resting-place, for dried and glossy stems of a circular figure accelerate, more than they retard, the backward gravitation of the body.

The creature, therefore (unable to reason, acting under the impression that its body is continually sliding backward), endeavours to recover its original position by nudging itself repeatedly forward. The horse has neither light to see, hands to feel, nor sense to measure the distance. Imagination is the only dependence which it can boast of. The advances become energetic in proportion as the supposition which provokes them is annoying. The annoyance is regulated by the irritability of the quadruped. Some stable inhabitants grow more morbidly nervous; with these, the advances are proportionably frequent; so that the head of the captive, guided by the collar-rope, is speedily brought into violent contact with the further end wall of its compartment.

Not comprehending the meaning of the blow, but suffering from pain and fright, the animal attempts to rise. The commencement of this movement, always, is the elevation of the head, which, after being raised, is strained backward. This action is a necessity of its existence; and, dreaming of no danger, the quadruped essays to fulfil the natural law. The head, however, which has struck the wall of the stable, must at the time be immediately under the manger. Imagining no impediment, the animal exalts its crest with that impetuosity

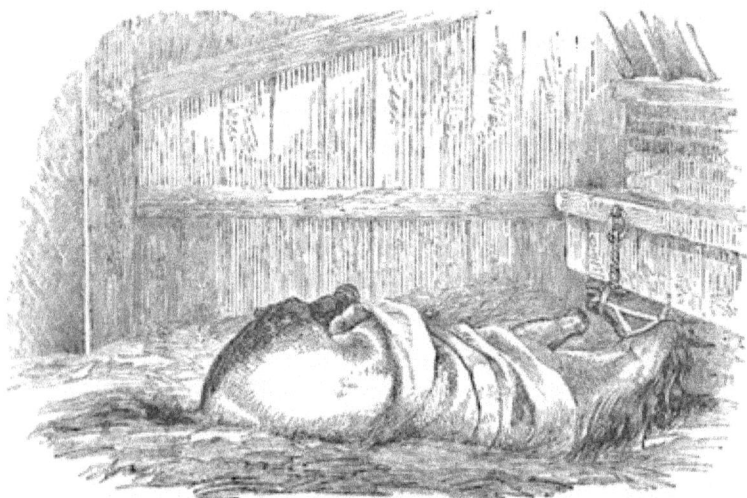

CAST UNDER THE MANGER.

which characterizes all the motions of the horse. It strikes against the manger, and a heavy concussion sends the member into its original abiding-place.

The stricken creature cannot comprehend the reason of those blows it has received. But it is often chastised for nothing, so beating is to it almost a matter of course. It crouches in terror for some moments, no doubt hoping its tormentor may move onward. Then, as the strained senses can detect no sound, it ventures once more to raise its head. The result is the same as it was before. The horse, after repeated efforts, becomes alarmed. Mad with fear, and wild with desperation, it now exerts its utmost strength. The contention may continue until the groom enters the stable in the morning, when, bruised and panting, its head swollen and bleeding, its strength exhausted and almost its life expended, the wretched animal is discovered prostrated upon the pavement.

This consequence of confining an animal in darkness is the serious and, probably, the permanent deterioration of property. At the best, its services are lost for many days. In any case, time must be allowed for the necessary recovery. Few, very few people have the generosity to recognize, and even fewer still are educated to perceive, that a life has been for many hours breathing in agony, and that the existence may hereafter, notwithstanding all the present state of art can accomplish, probably drag its wretchedness about the world in a crippled condition. No person living possibly will, when inspecting the maimed and disabled horse, reflect upon the fate which dooms the animal to years of sorrow, labouring through the lowest species of earthly trial; no one will heave a sigh that such a fate overtook a placid, gentle, and obedient creature, which was dangerously and cruelly confined during the time of serving a being who was bound to study the necessities and administer to the happiness of the life over which he had assumed absolute authority.

Other evils, also, spring from obliging the horse to sleep on a surface which is not level. The head of the animal being fastened to the manger, it has no choice but to couch where it stands, or to remain erect and endeavour to sleep in that position. There are quadrupeds which adopt and which maintain the last alternative; their bodies never repose on earth until their injuries and their wrong are engulfed in the common doom.

It is not every animal, however, which can hold to such a resolution, in spite of the aches and agonies by which it must be enforced. Certain creatures, feeling their bodies glide backward, rather facilitate than endeavour to counteract the motion, hoping to soon rest upon the gangway, which experience has taught them terminates the stall. Others sleep so soundly as to be unconscious of the movement; while a third class,

A HORSE STANDING WHILE IT SLEEPS.

having attained philosophy through a life of misfortune, pay but little regard to the circumstances around them. In all instances the frame descends the slope till the quarters pass the gutter, and repose upon the gangway.

Yet, before the body can move such a distance from the manger, the neck and the collar-rope must both be strained. However, finding its body, at length, to be comfortably located,

the animal meditates composing itself to sleep, which is not to
be done while the neck is outstretched, and the chin is raised
far above its natural position. To accomplish this, the muzzle
must be considerably lowered, and the neck be retracted; but,
before either can be done, the collar-rope must be loosened.
It is obviously impossible to change the attitude while that
fastening remains in a state of tension; the position in which
the horse invariably sleeps cannot, therefore, be assumed.

In this dilemma, the intelligent quadruped determines to
rise and to return to the manger. But a natural law has
ordained, that before the horse gets up from the ground the
head shall be thrown backward; thus lightening the weight
upon the fore quarters, which parts are always first raised.
The straightening of the front limbs is thereby facilitated.
But this movement cannot now be put in practice because of
the rope which retains the neck outstretched. Struggles are
useless; the position is fixed, and the creature is powerless to
alter it. The limbs are free, but these can only be used to
kick and to aggravate the pain of the situation. The animal
is a prisoner, and so it must remain, vainly contending with
its doom, and exhausting its energies in fruitless efforts to
escape.

Assuredly he should have possessed an enlarged capacity for
evil who first conceived the notion of making a living creature,
which was conspicuous for its strength, its activity, and its
timidity exist in a niche; to have its head tied up by day and
by night, and subsequently doomed it to rest upon a floor
which sloped in a painful and an unnatural direction. No
surer means could have been invented of shortening the life,
of deforming the body, or of injuring the limbs of the creature
in whose prosperity man conceived he had "a property."
Arms of all kinds, and of every description, the quadruped
might have been safely trusted with; but to require of activity
that it should be fettered and forego all motion; to demand of

timidity that it should be bound or imprisoned, and not display sensibility; to ask that strength should endure, and not attempt to struggle, was surely expecting too much from an inhabitant of a world in which fear, as the natural instructor of organized beings, is universally prevalent.

The horse, thus located, was only presented with the ready means of doing injury to itself. It was provided with the only weapons which Nature had empowered it to employ. A more unwholesome, a more unnatural, or a more dangerous abode for any of the equine race than the stall of a modern stable, it would be impossible for the utmost stretch of the most excited malignity to imagine. Still, daily accidents, which must have occurred for centuries, seem to be incapable of instructing mankind, where the welfare of another and of an inferior being is concerned !

Animals have been lamed; have lost the power of vision; have bred terrible disorders, and have been found stretched to death upon the straw bed, in consequence of the folly which has persisted in building modern stables. Such accidents must, as a necessity, continue so long as these edifices are erected. They are totally unsuited for the creature which they torture, cripple, and confine. Yet, because such abominations are sanctioned by custom and approved by ignorance, it is far more than probable that the author's exposure of their unfitness will be read with amusement, and admitted to be just; but the scourge which is recommended by its existence and patronized for its convenience, will still be perpetuated. It may continue to disgrace this country for more than another century; although the judicious outlay of a few shillings would greatly amend even modern stables. Banish the stalls, and divide the interior into loose boxes. Lower the mangers and the hay-racks to the floor, and abolish the loft, now placed over where the animals repose. Allow the entire space, from the ground to the roof, for the huge lungs to breathe in. Improve the drainage. Warm the building, by means of a slow combustion, and by water-pipes. To effect all this should not cost very much; and, as his reward, man would gain the longer service of his slave, together with an inward approval, springing from a consciousness of having done his duty towards the meekness which has been entrusted to his careful keeping.

CHAPTER VIII.

THE SO-CALLED " INCAPACITATING VICES" WHICH ARE THE RESULTS OF INJURY OR OF DISEASE.

MANY injuries occur to horses which incapacitate them from performing physical exertion, and notably among these is rick or chink of the back, which is among the most common and the least understood of equine affections. Its symptoms are confounded, one and the same name being employed to indicate every stage of the disorder; thus confusing inquiry. Those effects which result from organic change, are regarded as the promptings of that " viciousness of spirit" with which it has pleased mankind to credit the horse. The liberality of mortal imagination is extreme, especially where causes have to be assumed. Grant man the right to conjecture, and there is no mystery in Nature for which he cannot account. Thus, the sharp pangs of agony which induced the contortions of a dumb creature were conjectured to be the gratification of an innately " vicious disposition." This pretended explanation has remained unquestioned for ages, abusing the intellect of mankind, and hardening the hearts of those whom it was thought to enlighten. No doubt many very worthy people will feel much inclined to quarrel with the book which presumes to question the interpretation that generations have approved, and time has sanctified.

However, to expose the manner in which the personation of

meekness has been abused by the arrogance of ignorance—
certain animals are supposed to indulge a morbid habit, or
"vicious" propensity, which is, by the lower orders, spoken
of as " kidney dropping." Creatures thus viciously disposed
are generally aged, and are devoted either to heavy draught or
to harness purposes. They are sometimes met in those stables
where horses are let out by the " hour, day, or job." One
thus afflicted will be drawing a gig along some pleasant
country road when "the vice" shall be suddenly displayed.

"A KIDNEY DROPPER."

The attacks may appear in rapid succession, when they render
the life worthless; or they may only come on at distant
intervals, being separated by long periods of apparent soundness. No jockey, however knowing he may be in his vocation,
or however boastful he may be about " my 'sperience 'mong
'orses," can, by any visible sign, announce the day, or foretell
the hour, when a particular quadruped will be afflicted with an
attack of " kidney dropping."

15

The horse shall be harnessed to some light vehicle, within which may be seated some tradesman. The animal is not overloaded, and seems to be journeying pleasantly at his own pace, when suddenly the gig is brought sharply up, and the pony is discovered squatting upon its haunches like a dog.

This is an unnatural position with the horse. It is perfectly true animals are made to assume it in the circus of most amphitheatres; but if the reader remembers, he also beheld men in the same place put their arms and legs in positions which were quite as unnatural to humanity in general as sitting on their haunches possibly could be to the community of the equine race. What, therefore, may have been exhibited at a circus signifies nothing when regarded in its fitness for universal application; in all other spheres, sitting on the haunches, when exemplified by the horse, must be accepted as proof of bodily derangement.

If the attitude of the animal be observed, the hind limbs will be seen to have fallen in such positions as suggests no notion of comfort or of design. They may cross one another, or they may be sprawled out on either side of the body; they are never arranged with that grace and care which indicate the attitude to have been deliberately assumed. Moreover, should the skin be pricked with the point of a pin, no sign of sensibility is usually elicited from the hind quarters. Strike the prostrated members, and no evidence of pain follows the blow. The posterior portions of the body obviously are dead to this world and to its malice.

However, do not fuss about the horse; allow the sufferer to remain undisturbed where it has fallen. Have patience with the distress which no cruelty can quicken. Loosen the harness; remove the shafts; procure some water, and permit sensibility to allay its parching thirst. After a short space, the quadruped may get up of its own accord. Some time has been lost; but disease has not been aggravated by needless torture. When the creature rises, the fit has passed ; but the author doubts if the recovery can then be pronounced complete. He would certainly brave "an accident" who should essay to drive a horse but recently recovered from an attack of "kidney dropping," though this hazard may be frequently incurred with apparent impunity.

Allow the injured quadruped to remain in the stable undisturbed for the night. The following morning will be time enough for its examination ; for the disease under which the horse languishes is of a nature that cannot be affected by the lapse of a few hours.

The next day, having selected a piece of clear ground, cover the spot thickly with straw, and have the horse led on to it. The services of a veterinary surgeon are not imperative. The proprietor may himself conduct the investigation ; or, should he feel distrustful of his own ability, any person possessed of the necessary amount of confidence may undertake the active duty. All idle spectators should be first requested to retire. Then the investigator takes his position as close to the quadruped as possible. He runs the forefinger and thumb gently over the superior spinous processes of the vertebral chain, or down the centre of the back. This action is repeated several times, additional force being brought to bear with each succeeding trial, until the whole strength of the operator is exerted. While he is doing this, the person who undertakes the investigation fixes his attention on the head of the horse. If, upon pressure being made on any particular spot, the ears are laid upon the neck, or the crest is suddenly elevated, the fact must be mentally noted. The trial should be renewed, and if the like symptoms be elicited, the conclusion naturally is that the seat of injury lies immediately under or very near to the place indicated.

This point being ascertained, the operator puts a hand on either side of the tender part, and casts his full weight sud-

15 *

denly upon the spine. Such a proceeding, to be demonstrative, must be rapid and energetic. Horses under the sudden pang thus produced have shrieked in agony. Generally animals crouch under the torture, and burst forth into copious perspirations. The author knows of no instance where a desire to employ the teeth has been exhibited; although there is no predicating in what manner a creature may behave under the powerful wrench of actual torment. He, however, who undertakes such an inquiry must be pre-

TEST FOR KICK OF THE BACK.

pared for every eccentricity; and, while regretting the necessity which obliges agony to be inflicted on a gentle and timid creature, he should also be far above the coarse and brutal punishments which are too frequently indulged to check the writhings of the potent suffering.

The affair is thus decided. The spine has been injured, and the spinal cord which it sheaths is also involved in the lesion. Horses in such a condition are commonly, with an utter want of morality, cast upon the market, or publicly disposed of by auction. The animal sold is soon found to be worthless; it speedily passes from owner to owner, until it becomes the property of the lower class of horse-copers, to whom that which they call a "kidney dropper" becomes a prize: it is sold at a large figure, and is purchased at a small one again and again, until worn out with its sufferings death terminates the scene.

There is, however, another form of chink in the back, where the spinal marrow is in no vast degree involved, and in which the animal exhibiting the affection is not generally devoted to harness purposes. The horse is commonly showy in appearance, and is usually disposed of exclusively for saddle uses; but the existence of a disease is not denoted by any outward sign, therefore its presence is sneered at as a positive impossibility. Quadrupeds thus disordered are by the generality of horsemen condemned as "irreclaimably vicious."

One of the bones of the spine has been rendered loose in consequence of the ligaments being over-strained; the animal has been abused in some manner. The ligaments, when in this condition, are acutely painful; though no visual disorder may be observable to the post-mortem examiner, nevertheless, the slightest weakness in such a structure may, during life, occasion the severest agony. The bone is not fractured; but one of the vertebræ, through the leverage of its superior spinous process, may have been wrenched slightly to one side. This may not affect the appearance of the quadruped; neither may it elicit signs of pain when the weight is evenly seated upon the back; therefore only during the act of mounting, the drag then being entirely to one side occasions the most poignant anguish.

The horse, being dumb, of course cannot explain its sensations; nor can it appeal to the forbearance of its master. Its ailments are entirely subjected to the merciful consideration of man. The animal's actions, therefore, are always liable to

be misconstrued ; the promptings of torture are frequently
confounded with the exhibitions of the worst forms of "vice."
Thus a creature with the ligaments of the back strained is
always condemned as an inveterate kicker ; because the drag,
produced by the weight of the rider resting on one stirrup,
occasions so sharp an agony as alarms the quadruped, and
naturally excites a determination to repel some imaginary
enemy. The creature, consequently, commences to "lash
out" with its utmost energy. This violence is repeated so
often as the owner has occasion to remount. The action is
always sudden, and not to be inferred from the previous
aspect or behaviour of the nag. It is, therefore, attended
with the greater danger, not only to the proprietor, but also
to those who may be collected about the horse.

Violent, however, as may be the resistance provoked while
the foot is in the stirrup, the seat of the saddle is no sooner
attained than composure is restored. When the rider is once
fairly on the back, the steed assumes its natural timidity, its
docility, and its obedience. It is then transformed into all the
most fastidious proprietor could desire. That circumstance
has induced some horsemen, who were more thoughtful than
the generality of the race, to change the habits usual in this
country. Such persons have tried the effect of mounting upon
the wrong side ; this has usually, for a certain time, been
attended with perfect success ; but the custom, after a space,
has seemed to involve the sound ligaments, when the kicking
has been renewed with more than double vehemence. A horse
which kicks in the way described should always be transferred
to harness-work, when, no vast weight being upon the back,
the quadruped generally behaves admirably.

Rick, or chink in the back, is, however, the common property of creatures of heavy draught, and, with such a description of horse, the consequences are usually more marked and
much more severe.

The reader will readily imagine that a "kidney dropper," falling suddenly while pulling a weighty load, can hardly escape "accident." Therefore, quadrupeds of the coarser breed, and thus afflicted, rapidly come into the possession of those who do not scruple to trade with misery; and as this form of disease enables the sufferer to appear with a blooming coat, as well as a carcass carrying a quantity of fat, the copers often reap a rich harvest by their unscrupulous dishonesty.

Upon the earliest indication being perceived of the spine having been badly injured, the horse should be instantly thrown up for at least six months. The animal ought not to have a layer of pitch, rosin, &c., smeared thickly over the back, and be turned out to take its chance upon a green diet; but it should be placed in a roomy, loose box; it should have the hair cut off close over the seat of injury, and the place should be constantly moistened by means of cloths dipped in a lotion, composed of tincture of arnica two ounces, and water one pint. This remedy, with softened food of the most supporting kind, should constitute the treatment for the first month of recovery.

At the end of that period we may assume that inflammation has been subdued; thereupon, the measures adopted may be changed. Some compound soap liniment should be rubbed on the surface every morning. Should the application blister the skin, the liniment must be withheld for a time; but so soon as friction can be quietly endured, the stimulant must be renewed. All this while the quadruped should be well fed; but medicine should be strictly withheld, grass and bran mashes being solely employed to regulate the bowels if their action be sluggish.

When morbid sensibility no longer exists in the spine, and moderate pressure with the fingers can be borne upon the back, the liniment may be discontinued; but the restoration is to finish with the repeated use of liquid blisters. One side of the spine, near to the seat of injury, is first to be acted upon;

when the action of the vesicatory appears to be subsiding, the
other half of the back should be attacked. This plan must be
pursued till the fifth month has expired, the horse being sustained upon the best and most nutritive food. After this
period has elapsed, a handful of ground oak bark should be
mingled with each allowance of provender. The animal, during all this time, never being flurried, or allowed to leave its
ample stable.

Upon recovery, the quadruped ought never to be employed
for that same kind of service in which the injury was received.
No weight should subsequently be placed upon the back ; for
the spine, which has been once injured, can never by human
art be restored to its pristine soundness. However greatly
the animal may have been prized, even as a hunter, it is safer
and much more profitable to doom the steed to the collar, in
which last employment old hunters particularly delight in
exhibiting their highly-prized excellences of action. Many a
horse that appears in the London streets running before some
brougham, and which, by the gaiety of its spirit, excites the
admiration of the foot passengers, will, after death, be found
to have one or more bones of the spine joined by osseous
deposit, proving that the back, during life, must have suffered
serious injury.

Horse-owners, however, should be very careful not know-
ingly to risk chink or rick of the back ; for such an "accident," according to its intensity, may reduce the animal of
fabulous price to an article which shall literally be almost
valueless. It brings down the steed, which excited universal
envy, to the cripple, which no honest man would sell, and
which no prudent man would keep. The mischief, once established, too often sets science at defiance ; for the rick, when
bad, is terribly apt to terminate in fearful fracture of the spine.

Some horses, though not absolutely " ricked," are, nevertheless, stiff in the back. Such quadrupeds are unpleasant to

the rider, and are unable to turn in the stall; but whenever
their removal becomes imperative, they are backed out on to
the gangway, and then turned towards the door. A stiffened
spine can be no recommendation, but it may fairly be accepted
as evidence, that the animal has either been over-weighted or
has, in its time, done some hard work. It is invariably detrimental to the value: for the vertebræ being the base of the
anatomical body, their healthy condition is of the greatest
possible importance towards even an approach to soundness.

A POPULAR CURE FOR THE IMAGINARY VICE OF "JIBBING."

A horse which, upon hearing the command to proceed, will
commence throwing up its head, and, in spite of the whip,
shall back, is supposed to have learned the vice of jibbing.
On such occasions, however, various cruelties are commonly
perpetrated; but severity has then lost its power to quicken
timidity, because in five cases out of ten jibbing is the result
of a kind of equine epilepsy. Therefore, to resort to harsh

measures under such circumstances is only increasing the pain already inflicted by acute disease. Severity can only lend violence to the impulse, which is almost certain to succeed the attack. It may endanger the life of the driver, but it cannot shorten the duration of the fit. Every kind of brutality has been speculated in without effect. Such treatment, most probably, has prolonged insensibility; for noise, confusion, or agony is not likely to be sedative to the nervous system, which a word has morbidly excited. Yet, such practices are generally adopted. Nay, the author has heard of a professional man who, residing near London, possessed a fine animal which was thus afflicted. This person actually had some straw kindled under his quadruped's body, and, to quicken what he called "an obstinate vice," partially roasted the breathing flesh of his living property! So monstrous an artifice was successful on the first occasion; but, upon repetition, it ceased to operate. Such a custom is not unusual among the uneducated boors of distant villages; but the writer had hoped that no vexation could have induced an individual, possessing the most distant claim upon the name of gentleman, to adopt so inhuman and useless a resort.

The horse is a gentle creature; it has no courage; it can display no resolution. Its impulses always incline it to flee from danger. It is made up of alarms, and a child's puny hand may guide its huge strength. But the history of the animal supplies too many instances where the perversity of mankind has mistaken the prompting of disease for the display of malice. It is disgraceful to the boasted civilization of the present age that, while knowledge has much benefitted every sphere of human legislation, the errors, the practices, and the brutalities of the last century should be in full operation where the scant necessities of the most gentle, the most submissive, and the most valuable of man's earthly helpmates are concerned.

CHAPTER IX.

STABLES AS THEY SHOULD BE.

PERSONS about to build stables should, before they begin to plan, thoroughly comprehend the purposes which the new edifice is to serve; and in the matter before us, it is necessary that it be investigated on sound principles, scientific research, and well-weighed practical observation. This consideration, therefore, involves the necessity of the reasoner being acquainted with the habits of horses, whether placed in a natural or artificial state, and of his being versed in a knowledge of physiology and chemistry, to enable him efficiently to secure the adoption of hygienic measures in the construction of his stables.

In the first place, a site must be selected, which, if possible, should always be on rising ground, in order to insure the downward flow of drainage; and, secondly, the internal arrangements of the stable demand careful consideration, since stables, as they now exist, are tainted with the evils of antiquity. Loose boxes must supplant the place of the stall. Each box is to be 18 feet square; of these there are to be six, ranged in pairs, three upon either side of the interior. Every box shall be rendered dry and sweet by six deep gutters, three on either side, and all emptying into a central branch drain, which discharges its contents into a main drain, running through the length of the entire building.

The gutters commence eighteen inches from the side divi-

sions of the boxes : the first is situated three feet from the
external wall.   Six feet divides the first from the second
gutter; the same space separates the second from the third
gutter, which is removed only three from the central par-
tition.

The flooring or pavement between the gutters is arranged in
gentle undulations, like the walks in a gentleman's garden.   It
is raised three inches higher in the centre of each division, than
where its borders terminate in the gutter.   The two pieces of

MODES OF STANDING AFFORDED BY AN UNDULATED PAVEMENT.

pavement, at either end of the box, begin at the elevation of
three inches, and sink to the level of the lowest surface as they
approach the gutter.   Thus, every portion of the pavement
will incline one in twelve, a fall of fully sufficient magnitude to
allow of the speedy disappearance of fluid, which is always
ejected with force and in quantity.   The gutters all terminate
in "stink traps," which give admission into the branch drains,
these last, as well as the main drain, consisting of circular
earthen pipes.

The undulations of the pavement not only facilitate the speedy removal of fluid, and thus tend to keep in a state of purity the atmosphere within the building, but the surface presents every variety of standing-ground to the choice of the quadruped. The animal, by this arrangement, can select an upward slope, a downward incline, or a level plane, whereon to rest the foot: an ability of appropriation which intelligence will not be slow to comprehend or tardy to appreciate.

Each gutter should be two inches wide and two inches deep. They ought to commence at the depth of a Dutch clinker from

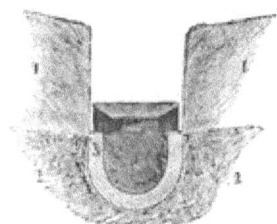

DIAGRAMMATIC SECTION OF A SUPERFICIAL GUTTER, SEVERAL OF WHICH KEEP DRY THE LOOSE BOXES.

1 1   The Dutch clinkers.

2 2   The prepared ground on which the gutters and the pavement repose.

3   The semicircular earthenware gutter along which the fluid flows, covered by the loose iron grating.

the surface, and be covered by a perforated loose iron grating, the holes in which are a quarter of an inch wide, one inch long, and the last distance asunder. Thus, should the horse, when down, lie over one of these gutters, the body cannot then repose on a good conductor of heat.

The gratings are not flat, but incline on every side towards the openings. This pattern was selected, because the author has beheld flat bars eaten into by the acridity of the fluid, and retaining liquid that yielded an abominable stench. Neither are these coverings fixed into their situations. They are merely laid upon the sides of the earthen gutters, which are three inches wide at the openings; the iron can afford to dispense with other fastening than its own weight supplies. Should the channel, which the grating guards, ever become clogged, then the easy lift of the metal-work will allow the gutter to be cleansed.

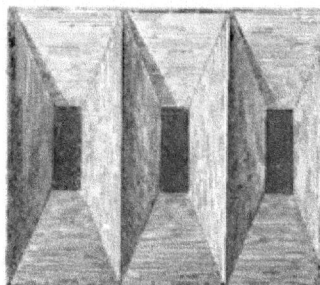

PATTERN OF THE LOOSE IRON GRATING WHICH COVERS THE GUTTERS.

The openings, which are ample to permit the escape of all liquid, are purposely made small, because rats and other vermin too frequently enter stables by the drains. It is by no means unusual for such pests, where they are numerous, to attack and gnaw the hoofs of living animals. The horn is without sensation, therefore it can be gradually removed without the horse being at all inconvenienced; but, assuredly, the proprietor will be vexed at a destruction which necessitates the quadruped should be idle until Nature has repaired the loss of substance.

The branch drains, which commence at twenty inches from the surface, can only be entered through a stink-trap; that article also opposes an obstacle to the free passage of vermin. All these branches terminate in the main drain, which, where the tube begins, is situated thirty-four inches within the soil, and, as it proceeds, has a fall of about one foot in fifteen feet.

Neither the pipes, the gutter, or the clinkers are placed within, or rest upon, unprepared soil. Such may be the usual plan after which most stables are now built; for the drainage of these places does not generally extend beneath the surface. The pavement of the contemplated stable, however, is to be raised two feet above the level of the ground on which it is erected. For the entire space which the structure will occupy the soil is, in the first instance, to be removed to the depth of one foot. After the foundations have been properly laid, the walls are then to be raised till they are built up two feet above the natural level of the surrounding surface.

A layer of large flints, or of coarse brick rubbish, is then to be thrown in; this layer is to be two feet six inches in thickness. Within this the main and the branch drains are to be arranged, though the principal drain will also have, towards its termination, to be sunk into the earth. The remaining six inches is to be filled in with coarse sand; upon this the gutters are to commence.

The gutters are two inches deep. They all originate at five inches from the upper surface of the clinkers. The shallowest has a fall of fifteen inches, but others have a much greater inclination; as all empty into the branch drains, which communicate with the main drain. This last, sinking deeper as it proceeds, quits the building at a depth of six feet six inches from the exterior of the sand within the walls of the stable.

The contemplated structure will be thus thrice drained. First, there will be the deep tubular main and branch drains; next, there is the sand and brick rubbish; while, lastly, there is the surface drainage, effected by the grated gutters. So much pains have been consciously bestowed upon the dryness of the building, because nothing will, in the end, prove more detrimental to the horse than confinement in a damp abode. Not only does perfect drainage conserve the health of the equine inhabitants, but it likewise tends to preserve the bricks, the mortar, and the expensive fittings that should adorn every stable.

According to the supposed view, which forms the frontis-piece to the present volume, there is a free but covered space, twelve feet wide, extending all round the building. The soil of this free space, covered ride or ambulatory, should also have been removed, and subsequently have been filled up, after the plan already described, as necessary for the interior of the stables. It need not, however, be paved with clinkers, as sand forms a better ground for a horse to exercise upon, than can possibly be made with the hardest of known bricks.

The roof, having sheltered the ride, terminates immediately over a metal gutter. This gutter communicates with five pipes upon the western and upon the eastern sides, with two pipes upon the southern, and with three upon the northern aspects of the building.

The roofing of the ambulatory is upheld by thirty-one posts, each twelve feet high, and the same distance apart. Between every two of these posts, on all sides of the stable save the front, are placed smaller uprights, which reach only to six feet. By these smaller posts are supported one end of three move-able bales on either side, the opposite extremities of the bales resting against the larger posts; each bale being six feet long, and reaching from the small uprights to the main supports. The first bale is one foot from the ground; while the others are at equal distances, and so placed as to leave four inches of clear pole to project upon the highest rail.

The pipes leading from the highest gutter are fastened to the pillars and empty into a drain, which encircles the build-ing and receives the water from the roof; it also conveys away that which is used in washing the carriages, or for general purposes. This is carried to any convenient pond, while the liquid manure of the stable is, by the tubular pipes, conveyed into a tank at least twenty yards from the principal building.

The bars forming the upper portion of the divisions are not so close nor so bulky, but the interspaces will allow the horses, after the Australian mode of cementing friendships, "to rub noses," or to exchange large draughts of fragrant breath with their fellow captives. Such innocent familiarities will often lead to lasting friendships, from the establishment of which the proprietor will reap an advantage. Quadrupeds perform much more gaily when harnessed with a companion that they love; and should the owner be at any time pressed for room, one or two additional spare boxes can always be commanded by allowing equine friends to enjoy the same compartment.

There is, however, running throughout society, a strange prejudice against permitting any communication between the inhabitants of the stable. Such a dislike cannot be justified by appealing to nature, as horses, when free to exercise a choice, always congregate in herds. Neither is it warranted by universal custom. In cavalry stables the quadrupeds are merely separated by bales, or by poles suspended at either extremity by chains, and hanging between the animals. The habit, also, does not gain any support from consistency of conduct: since the gentleman who will shudder at the possibility of any communion in his stable, will, nevertheless, allow numerous equine creatures to assemble together, and leave them without check, when he turns his stud into the field to be "freshened up" by a "run at grass."

The boxes have each a distinct entrance. The doors are in the wall, and open upon an ambulatory. Each entrance is nine feet high, and six feet six inches wide; all sharp edges and projecting iron-work, as hinges, latches, locks, &c., being strictly forbidden. Such things often injure animals, while in the act of passing through these openings, and should never be permitted to project in any well-managed establishment.

The folding doors are divided into two parts, though not

16

absolutely in the centre, since the lower portion extends only four feet from the ground. The upper part can be thrown wide, without releasing the quadruped. The ventilation is thereby rendered far purer, while the captive is indulged with a more animated view than the walls of the interior can afford. The quadrupeds will protrude their heads through such spaces, and remain in that position for successive hours, looking the pictures of mild contentment, and contemplating liberty, which a generous nature appears to have relinquished almost without

A HORSE LOOKING THROUGH THE HALF-OPENED STABLE DOOR.

regret. A simple creature may here in shade enjoy the summer breeze, as it blows aside the forelock; for if man is, by his position, forced to confine the steed, he is not compelled to aggravate the sufferings which necessarily attend the condition of captivity.

The doorway, being of those dimensions which have been

already described, should afford all necessary security, especially when the groom adopts the proper method of conducting an animal through the ample space.

No possible accident should impress the memory of the captive with the notion that doors and anguish are associated one with the other. The habit of the animal, being accustomed to advance the head through the upper space, would, moreover, be of some service in dispelling all idea of pain, should the impression have been received prior to the horse coming into the possession of its present owner. The sight, also, of the man, to whom the affectionate creature may be attached, would, moreover, attract the notice, and inspire the confidence of timidity.

The lower division of the door should, on fine nights, after dusk, be opened; that the prisoner may stretch its limbs, and bathe its hoof in the evening dew. So the grass is kept sufficiently short, not to afford more than a nibble, no harm, but much good, will arise from sanctioning so innocent a luxury as a stroll in the free air. The eye of the horse fits the creature to roam by night; and man should, by this time, have suffered enough to cause a doubt as to the wisdom of crossing Nature in her many wonderful provisions for the welfare of her children.

Such a suggestion may startle the prejudices which are inherent in the proprietors of most training stables. These places are, however, chiefly situated on the open downs, where ground is cheap, and the herbage scarcely affords a bite for the close-feeding sheep. Half an acre of such land could, without much expense, be attached to each box. On to this the flock might be turned by day; but so much liberty could be afforded the equine captive during the night. The racer being reared for speed, it is surely wrong to cramp its limbs by too stringent a confinement!

Within the loose box there is no rack for hay to strain the

16 *

horses' necks, and shake seeds into their eyes, which must be
open to direct the teeth.   The ordinary manger is also absent.
The horse does not sit to eat, nor can it lift the food to the
mouth; but, naturally, it lowers the head to its gratification,
and thus has no need to be accommodated with exalted fix-
tures.   As it can with ease feed off the ground, why should
man, in the nineteenth century, persist in forcing the animal,
which he domesticates, to forego the habits which Nature has
engrafted on its existence?

No rope fastens an animal directly under the opening to a
dirty hay-loft.   No puffs of cold wind, therefore, can blow
upon the quadruped through such an aperture; which is not a
loss, for horses are very susceptible to colds, which modern
stables are ingeniously arranged to encourage.   Like all life,
when hotly and impurely enclosed, the steeds become morbidly
delicate: the pampered daughters of the wealthy cannot possibly be more vulnerable to evil influences than are those
equine slaves, whose service demands a body vigorous with
health, strong, and able to encounter all the seasons in their
vicissitudes.

As to the supply of liquid, some arrangement is also needed:
the bottom of the water-trough is level with the surrounding
pavement.   The supply-pipe is commanded by a tap, and all
the receptacles can be simultaneously filled by means of the
tube that rises above the superior margin of the trough.
Below the earth is a conduit, which conveys away the superabundant liquid.   Into this tube or drain two smaller pipes
empty, both of which arise from the interior of the receptacle.
The smallest pipe reaches almost to the topmost edge of the
compartment, and is simply intended to prevent the possibility
of an overflow.   The other and the larger tube is inserted into
the bottom of the trough, and the removal of a plug, which
commands the entrance, permits the contents of the trough to
flow through this pipe into the larger conduit below, which

empties its contents into the main tubular drain. By turning on the supply, which is derived from a cistern to be hereafter mentioned, and by also opening the waste-pipes, all the troughs can at any time be quickly cleansed.

The cistern is situated in the boiler-house, and is elevated several feet above the level of the stable. The boiler-house adjoins the boxes, and from the raised cistern springs the supply-pipe, which is carried underground through the stables. Water, however, will always rise to its own level; this property convinces us that the troughs will be speedily filled whenever the taps are turned. The taps, by which the flow is commanded, are both placed in the first box; and, by this arrangement, the animal can receive fresh water four times daily, without fluid being carried to the horse. The contents of the customary pails are too frequently spilt by careless grooms. The horse naturally thrives best in a dry abode. Besides, the drink, as in Nature, is always before the creature; for, if presented only at stated periods, the draught may be offered when desire does not require liquids; or it may be withheld when thirst is so powerful as to engender a disinclination for solid nourishment. Moreover, servants are not always attentive to their monotonous duties; and the animal, in consequence, may be denied a necessary supply of fluid.

The water-troughs are, moreover, recommended by further reasons. Horses are blest with acute senses; and everybody must have observed the animal blow upon, or rather smell, fluid before it partakes of the refreshment which it needs. The stable pails, generally, stand about; such things are exceedingly handy; and we need not be surprised if they are occasionally used for other than for cleanly purposes. The troughs, being fixed, are secured to one service; the pipes, emptying into the receptacles, prevent the purity of the supply from being tampered with. The above advantages are also associated with the ascertained fact, that the horse, with water constantly before it, drinks less than the animal to which the pail is brought only after hours of enforced abstinence have generated a raging thirst.

The roofs of stables in large towns must necessarily be composed of slates; but this material by no means forms the best protection to the inmates of the stable against the heat of the summer's sun or the winter's frost. Between the slates and the rafters of such roofs, some non-conductor of heat should be interposed, such as felt. This simple expedient would, to a great extent, protect those living within the stable from rapid changes of temperature.

In country districts, fortunately, stable-builders can be choosers, and a cheap material for the formation of the roof is always ready to hand. This consists of straw; there is no better roof than a thatched one, either for a house, stable, or any other location where animals are compelled to live. This simple covering forms a perfect protection against storm, heat, and cold, and consequently is the one above all others to be desired as a stable construction. But of whatever material the roof consists, it is necessary that openings be made in it, in order to insure ventilation—*i.e.* the egress of impure or spent gases, and the ingress of pure or atmospheric air. Many systems of ventilation have been tried, but none have succeeded so well as Muir's Four Courts Ventilator, which consists of one or more chimneys placed on the highest elevation of the roof. Each chimney should be a cylindrical tube, divided into two parts by a mid-partition or diaphragm, so arranged as to allow the escape of impure air on one side, and the ingress of pure on the other. By this means a continuous stream of fresh air is supplied, replacing the spent gases, or those unfit for respiration, and consequently injurious to health.

Having secured good drainage and effective ventilation, our task is only partly accomplished, since we must now select stable-fittings; and in doing so, let us give way to a little physiological reflection. The horse, in a state of nature, eats off the ground, and drinks with his mouth on a level with his feet. Such being the case, then, it must be antagonistic to the provision of nature to compel him to eat hay from a rack situated above his head, and corn from a manger placed opposite his chest.

In many modern stables, the inmate of the stable is so compelled to feed and drink; but this evil is easily obviated by lowering the rack, manger, and water-trough to the level of the horse's knees, at the same time causing the bottom of the manger, rack, &c., to descend to the ground level. By the adoption of this arrangement, the horse would eat and drink in the way Nature intended he should. The manger or mash-tub should possess a foot-guard, which will effectually prevent the horse throwing out his oats or mash whilst feeding; the water-pot also should be made, without a plug, to turn upon a pivot, by which means the unused water may be discharged, and the water-trough itself kept as clean as a china bowl, and the water always fresh. If oats happen to be in a trough with a plug in the bottom, they are retained, while the water passes off. Any one who has used the common water-pot with brass chain and plug, which frets the horse and soils the water, will see in this, although so simple, a wise arrangement.

By due attention to the wise laws which Nature indicates, and which in the above chapter an attempt has been made to explain, "Life may be prolonged, although death cannot ultimately be defeated."

CHAPTER X.

CLEANLINESS, quietude, and regularity should prevail
in every stable. Where one horse alone is kept, the
groom should be placed over a lad; for a stable cannot be well
managed by one pair of hands. The door of the building should
be unlocked punctually at six o'clock. The horse should be
inspected, to see that no mishap has occurred during the night;
after which, the animal at present receives the earliest feed of
corn, mixed with two pounds of clover hay, cut into chaff.
While this is being consumed, the night-clothes should be
removed; the unsoiled straw divided from the soiled bedding;
the clothes should be spread out to become perfectly dry;
the exposed body of the animal should be again thoroughly
inspected; stopping (when used) taken from the feet; the
water renewed; the feet looked to; the clinches of the nails,
which fasten on the shoes, should be felt; the unsoiled bed
heaped into one corner of the box; the day-clothes put on;
and those things generally attended to which are required to
give the place a smart appearance.

Seven o'clock. The day-clothes are either allowed to remain,
are changed for lighter sheets, or are entirely removed, according to the weather; the horse is bridled, and the animal is led
forth to one hour's exercise; the helper or the stable-boy
throws every outlet open; puts the bedding out to dry, if
requisite; washes the pavement; sluices the drains; cleans
the manger; allows a full stream of water to flow through the

troughs—getting the building sweet and ready by the expiration of the hour.

This morning exercise is, in London, often neglected; but it should always be strictly insisted on. Perhaps it were better were the animals at once conducted from the place in which they slept, and led through the air upon the first opening of the doors; after which they could return to sweetened apartments, with bodies refreshed and appetites stimulated by the morning breeze. Moreover, it is better to divide the exercise and the work by as long a period as possible; and the food must be more nutritive and wholesome when eaten in a clean apartment, than when devoured in a chamber reeking with the fumes of twelve hours' imprisonment. No fear need be felt concerning the delay, as the exercise is no more to the horse than is the early walk before breakfast, in which so many gentlemen indulge with advantage to their constitutions. During winter, however, the morning exercise is often delayed; and then is seldom given. The only legitimate excuse for the absence of such a necessity to health is the presence of a severe frost. Otherwise, winter and summer, the early walk should never be neglected.

Eight o'clock. The horse is brought in, and being stripped, the grooming commences before the body cools. This is performed outside in very warm weather, but within the stable when the day is either cold or wet. Hair-cloth, dandy and water-brush; hay whisp, sponge and comb, are only employed in this operation. The hair-cloth is used, save in cases of absolute necessity, instead of a currycomb; the other things are employed after the manner in which grooms are accustomed to use them.

The groom should always cleanse the body in the line of the hair. To ruffle this causes annoyance to the animal, and interferes with the beauty of its appearance. The daily renovation ought to commence with the head. On this part more time and patience should be lavished than are usually bestowed. The groom is not perfect in his duty until his office affords pleasure to the creature on which he operates. The ears are smoothed and made glossy with the hand. Then the fore-quarters are dressed; afterwards, the animal is turned round, and the other parts are attended to; but one agent is always fully used before the next is introduced. The openings having been sponged and the long hair combed, the toilet is then finished. This being done, the groom sees about his harness, &c., till nine o'clock.

To ascertain whether an animal has been properly groomed, inspect the roots of the mane. Should scurf appear, set the servant to remove it. Also, finger the body, which should communicate no thick and greasy soil to the hand. Grooms will assert it is impossible to prevent these effects; but if their labour cannot clear the coat, they must be either very ignorant or very idle. It is useless to dispute with an inferior. Tell him you insist upon your desires being accomplished, and you will only retain the man who can effect it.

Nine o'clock. The horse receives another feed, consisting of two and a half pounds of soaked peas or of soaked tares, one quart of soaked and crushed barley, with three pounds of clover hay cut into chaff, and also steeped. All soil is removed from the boxes; the groom then returns to finish his harness. Every piece is unbuckled and cleaned separately, and all metal articles polished, after the leather has been overlooked and renovated.

Ten o'clock. The man goes to the house for the day's orders; these obtained, he returns to the stable; he finishes the harness, and he cleans the carriage. The cushions should be removed and daily aired; in hot weather, in the sun; in wet, or during cold seasons, at the fire. This is done before the vehicle itself is attended to.

Twelve o'clock. The horse has another feed of oats, with two pounds of properly prepared chaff.

Two o'clock. The horse, when not required by the master or mistress, is led out for two hours' exercise. When its services are needed, the eyes, nostrils, &c., are sponged over; the mane and tail combed out; the coat is dried and smoothed; the exterior of the hoofs slightly glycerined; the feet and shoes specially noticed; then the saddle or harness is put on, and the animal is walked, not hurried, round to the front door. If the quadruped's services are not required, the last directions are unheeded.

Four or Five o'clock. When the horse returns, either from abroad or from exercise, the bed should have been littered down, and the body should be slightly dressed; the night-clothes should be ready; the animal be fed. When the horse is out late, the groom and the stable-boy should be up to receive it. Further instructions will hereafter be given concerning the treatment of the animal's possible condition when it is brought home at unseasonable hours.

At dusk a small light is ignited, and placed in a lantern.

At Ten o'clock the horse receives the last meal, which consists of the same ingredients as the twelve o'clock feed.

In the foregoing directions, only those things have been mentioned which require to be executed with regularity. Many small acts are, of course, not named. These are done between the more important duties. But, as a general division of the labour, a good groom should always make the horse the primary consideration. Thus, the fore part of the day is entirely spent upon the quadruped, upon the harness, and upon the vehicle; while the afternoon (where such an arrangement is possible) is devoted to the employer or to the stable, and to those small matters which always demand attention.

Horses, when in the field, will invariably be seen either resting or sleeping during the hot hours of the afternoon. The cool of the evening, consequently, would be a better time for enforcing exercise than the period when, according to

existing customs, it is generally administered. In private
establishments, however, many of the latter proposals
would be attended with inconvenience; but there is no household in which the ten o'clock feed and the evening exercise
might not be undertaken, and, in several public com-
panies, everything here suggested could be accomplished.
The morning's exercise should likewise be given before the
day becomes hot or the light is fully confirmed. Then, the
quadruped is braced by the spirit of the hour, not rendered
miserable by the heat and annoyed by the stings of innumer-
able insects.

The only peculiarity in the above regulations consists in
the length of time over which the feeding and the exercising
are distributed. In the first place, the horse is, by Nature,
formed to enjoy the night much more than it is made capable
of roaming during the day. In the second place, the author
never dissected the carcass of an aged animal without finding
the capacity of the stomach morbidly enlarged, and the walls
of the viscus rendered dangerously thin by repeated distension. The manner in which the small digestive bag of the
quadruped must be overloaded, by the usual plan of cram-
ming five full meals into twelve hours, accounts for the latter
characteristic, and also explains why indigestion should rank
among the most fearful and the commonest maladies which
attend upon domestication.

The currycomb should be seldom used; but the generality
of grooms require to be cautioned concerning the use of the
whisp and the brush. The first article is generally brought
down upon the sides with a succession of heavy blows. Now
beating is not cleaning; neither is one act necessary to the
proper performance of the other. The brush is often applied
so quickly and sharply as to cause the animal to shrink. The
groom would not admire being himself dressed according to
such a method. The hair-cloth should be used to remove

impurities; the brush is employed to expel loose particles, and to smooth any hair which the previous process may have disturbed, or roughened; the whisp is intended to polish the coat. Any violence over and above that requisite to fulfil such intentions is needless cruelty, and should, when detected, be immediately checked.

The more important portion of a groom's duty, however, concerns the treatment necessary for a wet, a tired, a dirty, or a heated horse. Most servants are successful in dressing an animal when the stable is entered in the morning, but few comprehend how to groom a steed in any of the conditions which have just been named; and of that number fewer still care to stay out of their beds to cleanse the soiled coats of the creatures entrusted to their custody.

Clipping and singeing are processes which all stable-men greatly admire. However, before the grounds of their admiration are criticised, it may be as well to reason a little upon what appears to be a growing custom. British horses are deprived of the thick, warm covering which Nature bestows only in the winter. It certainly does sound somewhat paradoxical, when it is stated that the English allow their quadrupeds to run about in full costume during the summer's heat, but take off every protection as wet, snow, and frost approach. Certainly, if extra covering is requisite at any period, man, by great coats, cloaks, mantles, overshoes, respirators, boas, and comforters, has declared that Christmas is the time for additional warm clothing. But the groom protests it is impossible to keep a wintry equine garment dry: he says, that when the creature has been made comfortable the previous evening, the coat is often found to be quite wet on the following morning.

Still, in some very cold climates, it is not unusual to wet the garments, for the purpose of confining the animal heat, or of preventing cuticular evaporation; therefore, the moisture of the skin may be ordained with a benevolent design. But granting all the groom can object to wintry perspirations, the body which perspires is confined in a stable, and an impure atmosphere can occasion a faintness which shall provoke a copious cuticular emission. At all events, man has, in his treatment of the horse, made such egregious blunders, that he ought to be careful how he presumes, in future, to differ from the ordinances of Nature.

To illustrate the effects produced by a thick, wet covering, and by a thin, wet envelope, let the author narrate the result of a very simple experiment, which the reader may without much trouble institute for himself. Obtain two bottles. Wrap one closely in several layers of calico; around the other fix only a single, tightly-fitting covering of the same fabric. Saturate the cloths of both bottles with water; also fill the interior of each with the same liquid. Renew the moisture to the two coverings as either becomes dry. After twelve hours, test the temperatures of the contents poured from either bottle. That from the thickly-covered (which may remain wet) vessel will be unchanged, or warmer for its confinement. That contained within the thinly-protected enclosure (which possibly shall be quite dry) will be cold, very cold,—so cold, that in warm climates water is thus rendered a refreshing draught. Nay, the hotter the medium to which the bottles have been exposed, the colder will be the temperature of the thinly-coated liquid.

Now, the stable is always a heated medium. The animal with a thick coat is represented by the vessel with a thick encasement, the contents of which are not chilled by the moisture which saturates its envelope. The clipped steed is represented by the bottle thinly enfolded, the liquid within which became cold. But, it may be urged, the clipped horse is never moist. Then perspiration must be checked, and fever must be present; for, during health, the pores of the skin are never inactive. Where the coat is removed, superficial perspiration, accompanied with constant evaporation, must always be taking place. Where the hair is thick, moisture naturally accumulates, because the covering prevents superficial evaporation, and thereby checks the operating cause of internal frigidity.

For the reasons explained by the above experiment, horses which have been clipped or singed are thereby rendered more susceptible to many terrible disorders. Any internal organ may be acutely attacked, because the perspiration has, by exposure of the skin, been thrown back upon the system. Numerous hunters (which animals are always clipped) fail at the beginning of the season from this cause. Nor can the author comprehend the purpose served by the prevailing custom, excepting the propitiation of a servant's humour. It is said the animal moves so much more nimbly after the long coat has been removed. This may be the fact, though the author has hitherto seen no such marked change follow the operation as will allow him to deliberately corroborate the general assertion.

Moreover, let the servant, when he notices the animal for the first time in the morning, observe the breathing of the quadrupeds. The building has been closely shut for the entire night, and the impure atmosphere will necessarily excite the respiration. Now, it may not be exactly in accordance with the groom's notions, but scientific men have long known the skin and the lungs to be joined in one and the same function. Then what right has ignorance to expect one to be idle when the other is oppressed?

Perspiration only implies cuticular activity. It is a healthy action; the emission of the horse is only an effort of Nature to cast off those impurities which man obliges his prisoner to inhale. The clipped animal must also perspire, if it also inhabit the building, and remain free from disease. Its skin

must equally exhale, as a law of its existence; but the hair
being short, and the surface of the body exposed, the heated
medium in which the creature stands may cause the moisture
to evaporate as rapidly as it is emitted. Still, all this will not
satisfy the stable-man. It is not only the wetness of the coat
which he dreads, but it is the presence of dirt that he abo-
minates. Long hair attracts and protects mud, which, however, is easily removed from any substance, after it has been
allowed to yield up its component moisture.

Viewing the insensible perspiration as an established fact,
the prevailing customs are not unattended with danger. The
advent of the summer's covering is delayed, and the system
seems to suffer greatly during the subsequent period of changing
the coat. The pace flags; the spirits fail; and the quadruped
becomes more susceptible to disease at a time of year when
equine disorders are commonly more general and more virulent.

Yet it may be urged that in the winter season the roads are
far dirtier, and the long coat is so much more retentive and
more difficult to cleanse. Here again the argument returns to
the groom. It is true, a long-haired heel should not be made
clean after the usual fashion. The man should not take the
horse outside into the night air, and should not tie its head to
the stable walls. He should not dash a pail or two of cold
water over the soiled and heated members, and should not
lead the horse back to its stall, retiring to bed with a comfortable conviction that he has done his duty.

Whether it meets with equal favour from the life whose
heels have to sustain the deluge, no one has hitherto been weak
enough to inquire. That Nature intentionally clothed the
horse's heels with long hair, to keep lowly-organized parts
warm and free from dust, is a fact neither thought of nor cared
about. The man, specially retained to look after the quadrupeds, cuts away the provision which was instituted by the
Source of all mercy; then applies cold water to the organs

which Wisdom saw reason to shelter, leaving the members to chill and chap, while he retires to his repose.

The animal, with its dripping heels, is hastily fastened in a stall. The clipped legs of a horse are admirably adapted to exemplify the effects of evaporation. That portion of the body where the circulation is most feeble has to endure the

THE USUAL METHOD OF CLEANSING A HORSE'S HEELS.

effects of the process which can generate cold, even during the extremity of the summer's heat. Cracked heels, grease, &c., &c. (see "Illustrated Horse Doctor"), are the immediate results; and the master who makes the welfare of his steed subservient to the idle prejudices of his groom, is fitly punished in the lengthened period of his animal's compulsory idleness, appropriately finished by the payment of a long bill to the

17

veterinary surgeon. It is thought by many horsemen that all horses' legs should be permitted to retain the adornments which were sent by the bounty of Nature for the comfort of her creatures. The clipped or singed horse is a deformity; the colour is unnatural; the coat is dull and stubborn, looking most unlike that polished surface which is native to the beautiful quadruped. Moreover, those who live in a temperate climate should be content to forego certain elegancies which are natural to warmer regions; or, if they will have tropical loveliness, they should encourage it by those means which enable oranges to ripen in England, and not descend to meanness, which may expose their desires, but can deceive no one—not even the most ignorant in horse-flesh.

Supposing a horse to be brought home with unclipped, but with soiled heels; with the lower part of the abdomen covered by dirt, and the coat drenched with rain:—the animal is led into the stable; the bridle and saddle are removed; the body is first quickly scraped; then it is rubbed over with a few dry whisps; afterwards, it is lightly hooded and covered with an ample sheet. The master, who has hastily taken off his boots, and changed so much of his clothing as was wet, now returns, bringing a quart of warm beer in a pudding-dish, and he remains to see the quadruped drain the draught.

Horses soon learn to drink and enjoy malt liquor. Were such stimulants equally at their command, certainly the animals would excel their superiors in habits of intoxication. The majority of quadrupeds may, with the first few draughts, require a little coaxing; but the primary disinclination overcome, the craving for such an indulgence seems to be immoderate. An occasional stimulant is, however, very useful in the stable; it revives exhaustion, and restores vigour to the circulation. The timely administration of a quart of fermented liquor to a jaded steed has often prevented those evils which usually attend upon bodily prostration.

The drink being swallowed, the sheet is taken off, and the body made thoroughly dry with whisps and cloths. The lad again employs the scraper; the man, with a cloth, dries the eyes, channel between the thighs, chest, and abdomen, always performing his duties with gentleness, and discarding the cloth for a hay whisp where the hair is thick, or wherever the water appears to have lodged. While this is doing, the proprietor should comb out the tail, the forelock, and the mane; he should also discharge those many little offices which are not laborious,

CLEANSING AN EXHAUSTED HORSE.

but which add greatly to the comfort of a tired animal. Other portions of this matter will be treated of in another part of the present article—such portions being the food proper subsequent to fatigue, and the right method of cleaning the heels. However, it may be necessary to observe in this place, that before the quadruped is left for the night, the sheet should be removed and the usual night-rug put on the body.

When a horse is brought in covered with perspiration, it is

17 *

led at once into the stable; master, man, and boy should join
in its purification.   The lad takes the scraper, and, beginning
at the quarters, hastily presses out the excess of moisture,
while the groom procures a pail of cold and a pail of warm
water.   All being ready, the master not having left the stable,
the lad brings forth a dish of diluted soap (half a pound of
soap whisked about till it has dissolved in one quart of water)
and, dipping his right hand into this preparation, smears it all

CLEANSING AND COOLING A TIRED HORSE.

over the body.   So fast as the youth rubs the soap into the
hair, the groom washes it off, by pouring warm water over the
place.   The warm water carries away the soap, and with it are
also removed all the impurities natural to the soiled condition
of the skin.

After the groom comes the master, who pours upon the

body, already washed with warm fluid, a stream of cold water, from the rose of a watering-pot. The intention of the process may be thus explained:—The dissolved soap and the warm water are simply used to cleanse the body; having done this, the cold water is applied merely to close the pores of the skin, and to invigorate the system which exertion had debilitated.

This accomplished, all hands present, after the manner already directed, should set to work scraping, rubbing, combing, and using their utmost endeavours to dry the animal as quickly as possible. The horse is then lightly hooded and clothed.

DRYING THE HEELS.

With regard to the legs and feet of the animal, these parts are so much exposed that to them the same danger does not attend the presence of damp, as is commonly dreaded in the human subject. The water, with which the body has been drenched will naturally flow down the legs, and remove from them no inconsiderable quantity of soil. All, however, having been performed as directed, the groom takes up each hoof and

cleans it thoroughly out with a picker and a hard brush. Then
he goes upon his knees, and with several straw whisps he re-
moves as much dirt and moisture as will yield to friction. This
done, he brushes over the outer wall of the horn with glycerine,
and rolls bandages round the legs.

In the foregoing illustration, the size of the horse-cloth cannot otherwise than have appeared strange to the reader. But
things as large, if not of a greater magnitude, should be in
every stable—not for general use, but for special occasions. The
ordinary rug merely covers the spine, not doing so much towards
keeping warm the carcass of the horse as would be effected by
a Guernsey jacket upon the body of a man. Yet, who would
think of employing the last article as a sole envelope for a cold
and fatigued traveller? This, however, is all modern custom
sanctions for the comfort of a tired and exhausted steed! The
folly of so inadequate a provision is apparent, and the necessity of the innovation suggested by the last engraving must
be obvious to all who will condescend to think seriously on the
subject.

While the legs are being attended to, the supper may also
be before the horse. The meal, however, should not be of the
full quantity, or of a heavy nature. The stomach sympathises
with the general exhaustion of the body; the digestion is too
much weakened to appropriate its ordinary nutriment. For a
steed whose feeding capabilities are not hearty, a little bread
and salt, offered from the hand of its human favourite, will
frequently be eaten. Half of a half-quartern loaf, lightly seasoned, commonly will be gratefully accepted, if given in the
manner directed. Often, however, the craving is limited to
liquids, all solid provender being refused.

The animal should not be annoyed by any well-intentioned
coaxing to eat, when Nature commands it to abstain. The
inclination of the quadruped should at this time always be
respected; for a tired steed stands upon the borders of inflam-

mation, and in proportion to the value of the quadruped invariably is the danger of an attack. Hard-worked horses often want the stamina which enables nature to resist the effects of exhaustion. The bread, if not accepted, should be immediately withdrawn, and a pail of well- and smoothly-made gruel, with which the meal was to have concluded, be alone presented. All other food should be removed, and the animal left supperless to its repose.

GET UP!

If the gruel is rejected, take it away; place it in a cool situation, and it may be swallowed with avidity on the following morning. If allowed to remain, the animal will breathe upon it, and grow to distaste the nourishment. Suffer the horse to take the rest, which a disinclination to feed will have informed you is nature's primary requirement. Only, order the groom once or twice to peep at the nag through the

window which overlooks the stable. Should the creature
have lain down, the man may retire to his bed, convinced that
all is well; but should the animal, upon second inspection, be
beheld standing up, no time must be lost. The servant ought
to dress himself, to apprise his master, and to descend to the
stable; for this attitude, being long maintained, is among the
earliest and surest indications that disease has commenced.

A good feeder may simply require an allowance of bruised
beans and corn, to be well boiled in a sufficiency of water,
and before being presented as two meals, quite cold. No hay,
but a little bran or chaff should accompany the mess; as the
desire is to nourish the system, without overloading the
stomach. Should, however, this potion be refused, it is soon
converted into gruel, by stirring to it a sufficiency of water,
and placing it on the fire; afterwards by pouring the liquid
through a strainer, the husks are readily separated. It is
but seldom that full feeders are thus far exhausted. A vora-
cious appetite is, commonly, united to so much slothfulness of
body, as saves the horse from the aggravated effects of absolute muscular and nervous prostration.

On the following morning (supposing no mishap to have
occurred), when the time arrives to groom the horse, the
bandages should be taken off, and, as each wrapper is removed, the leg ought to be dressed. Firstly, the member
should be well rubbed with several whisps of straw. The
more apparent dirt being removed, the part should be further
cleansed, by application of the hand. After this the hair should
be combed; then again ruffled with the hand;—these processes being terminated by a thorough application of the dry
water-brush. This operation should be repeated upon each
leg, no hurry being indulged in the performance of this operation; but water should not be applied to the heels, without
the special leave of the proprietor having been obtained.
The case should be very marked before such permission is

accorded, for a wet heel gives rise to many troublesome affections.

All grooms are much disposed to treat the foot of the horse as a mysterious organ, which none but a person reared in a stable possibly can comprehend. The foot is not generally understood; because people, in their folly, will insist on regarding a very simple member as an uncommon and a com-

THE GROOM, ON HIS KNEES, TAKING OFF THE BANDAGES AND RUBBING THE DIRT OUT OF THE HORSE'S LEGS.

plicated structure. The horn being porous, insensible perspiration should escape through its minute openings. To prove this, let the gentleman accompany his nag to the farrier's the next time the animal is shod. When the sole is pared, let a wine-glass be held over the part, and the surface of the vessel will speedily be bedewed with the exuding moisture.

Now, grooms understand nothing, and care less about the perspiratory property of the horn. They cannot understand how the stoppage of perspiration may induce serious sickness. Therefore, most of the secret nostrums employed to embellish and to keep healthy the horn of the horse's foot, contain tallow, wax, lampblack, and various solids, which must clog the pores of the hoof, and, by arresting one of its functions, provoke disease. The best application to adorn this part is a little of the glycerine mixture, directions for preparing which have already been given. This moistens and renders pliable the hoof, which, be it black or white, will present a polished surface, without the pores being clogged up by the tenacious property of its substance.

Let grooms who wish to understand their duties, and who are not above taking a few hints, avoid those stablemen who (according to their own thinking) possess a knowledge which masters impossibilities; which can make any lame horse go sound; can induce prime condition in less than a week; can cure glanders; can render the most savage horse as tame as a lap-dog; which knows how to plan a stable; how to make harness look well and last long; which understands carriages; and, in short, everything within and outside the stable. Such men lead the uneducated into error, and error which sometimes proves very injurious to the horse, and often ends in the dismissal of the servant. Let the groom, from his earliest moment of service, make up his mind to serve his master with truth and honesty, and to avoid the company of the reckless and ignorant, and to seek advice of those who are competent to give it; to treat with kindness and consideration the lives of the horses committed to his charge; to use kind words instead of harsh expressions; to educate the animals to fondle rather than to fear; and, in the long run, and short too, you will discover,

O, groom! that, with three weapons,—truth, decision, and kindness,—you will be able to direct horses almost with your hand, and to conquer any man who may chance to be your enemy.

"KIM OVARE."

CHAPTER XI.

THE skeleton is the bony framework of the horse, and the vertebræ are the base to which all other bones concentrate; and we notice that at one end of the vertebral column the skull is placed, and at the other the tail.

When the reader has been riding in any vehicle and looking down upon the spine of the horse, he can hardly have failed to remark, that the widest portion of the body was the prominence of the hip bones. The posterior parts, or those

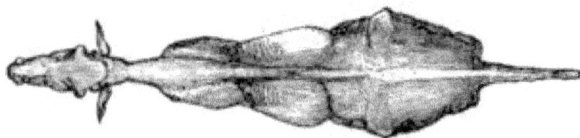

THE BACK OF THE HORSE, AS SEEN FROM ABOVE.

behind the projections, are not continuous of size; but they, nevertheless, are far more bulky and altogether more fleshy, than any of the forward surfaces of the body. Flesh is only another term for muscle; consequently, where flesh is most conspicuous, strength most resides. The muscles of the hind limbs spring from a large bone, variously named in common parlance, as the haunch bone or the pelvic bone. It is also spoken of by anatomists as the os innominatum. This large bone joins the spine at the hips, and thus lends support to the posterior region. But the vertebræ, immediately before the hips, are aided by no such accessory. The loins stand alone, or are placed entirely without support. This part of the body merely consists of certain bones, over which and under which run thick layers, or solid masses of muscular fibre.

A thorough comprehension of the osseous weakness apparent in the skeleton of the loins, must convince the reader of the absolute necessity which exists for some compensating agency, so as to fit the back for its burthen. The loins, therefore, should be bulky or muscular. They cannot be too large, but may easily be the reverse. Small loins or weakly loins admit of no compensation. The reviser does not remember an in-

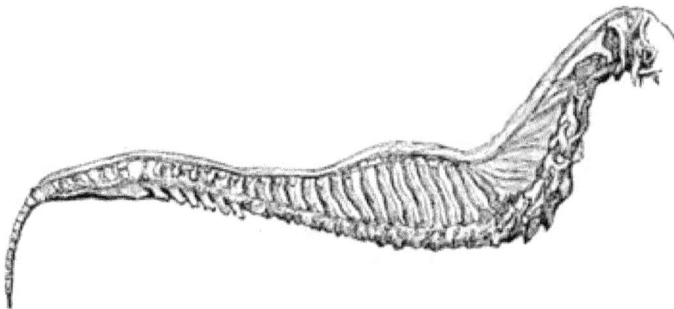

THE BACK BONE OF THE HORSE.

stance where such a formation was not associated with mean quarters; whereas, he does not recollect a case where size, in this region, was not evidence of general strength and of remarkable vigour. The position of the part is peculiar. It is intermediate, and lies between the haunches, which are the propelling powers, and the thorax, which region is formed to endure, to support, or to uphold what the back carries. All intermediate structures demand strength; because such a position exposes the part to the full impulse of adjacent force, its office being simply to transmit that impetus which it directly receives. Accordingly, the development of the loins,

both in man and in horse, may be cited as the best proof of
the vital power which resides within the frame.

The loins, to evidence the transmitting office peculiar to this
region, receive and convey onward the propelling force of the
quarters. So, when the body is suddenly checked, the loins
have to master the first energy of the onward impetus, or
have to endure the full violence of the sudden arrestation of
the forward motion in both the animal and its burthen. In
the brief, but dangerous, feats of leaping, galloping, &c., the
position of the region and the duties involved by it are so
obvious, that the author cannot presume to dilate upon what
appears to be self-evident.

Muscular loins are imperative in racers and in hunters.
They should, also, characterize all saddle-horses; for it is
impossible the rider should be safely carried, unless the back
be strong. The animal designed for light harness purposes
can, perhaps, best dispense with such an essential; although,
even in that case, the deficiency is very far from a recommendation; for weak loins are usually associated with a narrow
chest, a lanky frame, and a total lack of every property which
characterizes endurance.

In fact, every purchaser should first glance at this part; for
here reside those proofs which the scientific mind and the
practical judgment unite in esteeming. No matter what
quality may be desired, be it strength or appearance, be it
speed or endurance, breadth of loins is always important.
Lumbar development is essential in all cases. In short, there
is no property for the possession of which the quadruped can
be valued, that is not more or less, and generally much more
than in any degree less, dependent upon this portion of the
frame for its exhibition.

The backbone of the horse—lumbar bones and all—are often
remarkable for very opposite developments which pervade its
entire length. These are sinking down or curving inwards;

and rising up or arching outwards. When the line declines more than usual, the form is denominated a "hollow back" or "a saddle back," and is generally, supposed to be indicative of dorsal debility. Animals, of such a formation, however, commonly are possessed of high crests, of full loins, as well as lofty haunches, and they generally exhibit very proud action.

A HOLLOW-BACKED HORSE.

Animals with hollow backs are usually conspicuous, even among the equine race, for many estimable qualities. They are generally very docile, and uncommonly good-tempered. Putting the undue sinking of the spine out of the question, they display numerous excellent points; and, even admitting all that may be said about weakness, they exhibit such prominent good qualities as in many occupations may be justly esteemed more than an equivalent for their bodily deficiency, —especially when employed to carry a lady's saddle.

The very reverse of all that has been recorded above usually characterizes the "roach back." The author has, hitherto, found creatures thus made distinguished for the absence of

that power with which prejudice is inclined to invest them.
Such animals are to be seen feeding upon the commons about
Essex, being the pictures of checked development, and the
representatives of heartless neglect.   The offsprings of aged
dams, or colts that have been forced to submit to early labour,
every feature testifies to the abuse which they have undergone.
Quadrupeds, equally misshapen and equally neglected, may
frequently be seen dragging agricultural carts through the
streets of London.

A ROACH-BACKED HORSE.

Such deformities are usually vicious and spiteful.   They are
capable of little exertion; and offer a seat of torture to the
individual who is so unfortunate as to be mounted upon a
roach back.   Some years ago, the author chanced to dissect
the body of a quadruped of this description.   Death had not
affected the upper protrusion of the spine, which retained its
peculiar curve.   The loins were very poor, and several of the
lumbar bones were joined together by abnormal, osseous deposit.   The quarters were mean, the belly large, the withers

low, the neck ewe-shaped, the head big, and the legs long. In short, such horses are equally misshapen and mischievous. Any gentleman had better endure fatigue than accept such a creature for the companion of his journey.

Neither a long nor a short-backed horse is, necessarily, desirable. All depends upon the strength of those muscles which support the spine; though, all other points being equal, length generally provides a springy seat for the saddle; whereas, a short back commonly possesses the greater endurance. A long back, having bulging loins, is, however, infinitely to be preferred to a short back, with deficient lumbar muscles. The mere extent of a part can be no absolute proof in either direction; though, should a choice lie between two carcasses, supposing each to be equally deficient, or both to be equally favoured, then the short back should be preferred, because all increase of length necessitates a greater strain upon the organs of support.

But the spine cannot be too long, supposing length to be accompanied by a proportionate excess of muscle; for length and strength, of course, increase speed. The practice, common among the vulgar, of placing the open hand upon the upper part of the abdomen to ascertain the distance of the last rib from the hip bone, is a silly custom, and can prove nothing but the ignorance of those by whom it is exhibited. A living body should be judged as a whole. One part should be viewed in its relation to another development. No opinion on such a subject ought to be formed upon any solitary test or independent development.

The movements of the horse's body are greatly governed by the position of the head; it is therefore imperative, for the ease and safety of the rider or the driver, that the head should be well set on, and should be carried without sensible restriction. Should the rein be held too tight and a false step be made, or should the foot be placed upon a rolling stone, the quadruped

18

is almost certain to fall; for the rapid motion of the head
being impossible, it cannot be used to restore the disturbed
balance. The nimbleness which could avoid sudden danger is
destroyed by the fashionable want of feeling.   It is a matter
for surprise that the presence of the bearing-rein is never
alluded to, when gentlemen seek redress, because their vehicles
have been damaged. Most horsemen, however, esteem the
neck for its appearance, and few comprehend its utility.

Many judges would object to a bulky neck; but this form
indicates the presence of muscle, and a neck, if properly
shaped, cannot be too thick.   But some horses possess a bull
neck; this is almost confined to the heavier breed of cart-

A BULL NECK.

horses. It is, however, falsely denoted to be an evidence of
excessive strength.   So far as thickness is concerned, muscle
or adipose tissue must be present; but the length of neck is
deficient, which will limit the amount of motive power: in
fact, such malformed horses are known to be incapable of
severe exertion. On the other hand, a narrow and small neck,
although sometimes associated with low condition, is often

met with among well-bred horses : a horse with this form is
said to be ewe-necked. Such a form may possess length, but
it is wanting in depth and substance. Animals of this formation are generally active, but weakly ; other parts are too
often characterized by a narrowness of build which materially
detracts from a capability for endurance. The appearance is,

THE EWE NECK.

moreover, mean ; this is usually rendered more conspicuous by
a thinness and a shortness of mane. The shape of the neck is
not, however, to be considered only as governing other organs,
but is also to be regarded as a consequence of a prevailing
absence of development. So may the frequent accompaniment of a vicious disposition be viewed as the result of that
feebleness which converts the easiest task into a mighty
labour ; and of that absence of beauty, which can neither
kindle the pride nor awaken the fondness of the owner.

Certain supposed judges are greatly prejudiced in favour of
a short neck. The characteristic is in some minds associated
with the presence of bodily strength ; but it cannot be remarkable for denoting the existence of such a quality, because an
absence of length must abbreviate the amount of muscular

18*

fibre. Shortness of neck, besides suggesting the presence of
fat, and interfering with activity, unfits the animal for certain
situations. A bull neck, although its possessor inhabited the
most luxuriant pasture, would compel the creature to subsist
on short commons. *Nags*, however long may be their legs,
or short their necks, generally manage to crop the grass;

THE MANNER IN WHICH A SHORT-NECKED HORSE MANAGES TO FEED OFF THE GROUND.

although to do so may cause a constant strain upon their
limbs, thus counteracting one of those effects which the run
is invariably supposed to realize. The above illustration is
inserted to show the artifice adopted by animals of this description.

Having noticed those portions of the spinal column in which
the vertebræ are not associated with other bones, or do not
enter into the formation of compound parts, it may assist the
judgment of the reader if the relative importance of those
regions be more particularly descanted upon.

However desirable an arched and lofty crest may be, it is not, when separately considered, any absolute proof of estimable properties. Conjoined with other points, it renders excellence more excellent; but, alone, no deduction should be drawn from it. In many parts of Germany the horses exhibit beautifully-formed necks, bearing luxuriant manes; but in other respects the quadrupeds are lanky, weak, and washy creatures. The dock deserves attention, although it can warrant no more than an inference. If it suggests that which other developments equally support, it constitutes a valuable accessory towards a sound opinion; but by itself it is of no importance. On the contrary, the loins are absolute proof; their swelling testimony may be trusted, should both neck and tail oppose their evidence. This portion of the body never deceives. It is worthy of all reliance; what it declares must be implicitly received. And, to many minds, it may appear the more deserving of estimation, because full loins are commonly accompanied by a stout dock.

Attached to the neck is the head, which in the horse always bespeaks those changes produced by varieties of treatment and difference of climate. The favourite and the companion of the semi-civilized Arab is, by its association with its master, elevated in intelligence as in beauty. The agricultural teamster of this country exhibits in its expression the apathy with which it is regarded by its rustic attendant. These are, probably, the extremes of the race. That the reader may recognize the distinction between them, front views of both heads are annexed.

In the Arab, the spectator can hardly fail to remark the distance by which the eyes are divided. The brow is equally characterized by its length as by its breadth; and constitutes no mean portion of the entire head. In the lowly-bred face the region of the brain is comparatively small; its width presenting no obvious contrast to the other features. The nostrils are not only compressed, but their margins are thick; while the upper lip is adorned by a pair of abundant mustaches. Some animals the author has beheld with embellishments of this order which would not have disgraced the most hirsute of guardsmen.

A LOW-BRED HORSE.

A WELL-BRED HORSE.

The head of a well-bred horse has been frequently described as forming a straight line in its forward margin, when it is contemplated from the side. Such an assertion is generally true; but it must not be received as absolutely correct. Horses have been imported from Arabia with the craniums and the frontal sinuses considerably enlarged. Such a peculiarity is not esteemed a defect by the natives of the East. This fact is established by animals, thus characterized, having been sent to this country as presents for personages of exalted rank. Such developments may not strictly accord with English notions of equine beauty; but the size of the case, in some measure, denotes the magnitude of that which it contains. A large brain can be no detriment to any animal which is partly prized for its intelligence.

Another peculiarity exhibited by a few English thoroughbreds, is the Roman nose, or a prominence of the nasal bones.

The trait is, however, less common in the pure Arabian blood, than is the previous development. There is a breed of blood

BULGING FRONTAL SINUSES.

horses which exhibit a prominence of the nasal bones, or, in other words, present what is designated as the "Roman nose."

A CART-HORSE, WITH THE ROMAN NOSE.

This particular shape, however, is with the coarser breeds far from unusual; although, in animals of slow work, it cannot be esteemed a beauty, it also should not be condemned as a huge defect. The depression of the nasal point may allow less

freedom to the nostril; but in a creature, whose kind of labour
permits slow respiration to be employed, this constitutes no
absolute objection; while many quadrupeds of this formation
are conspicuous for their high courage and their lively dispo-
sitions.

The leading or distinguishing characteristic of the thoroughbred horse is its superior intelligence. The stranger hardly
has spoken to the creature, before it begins to investigate his
personal appearance. It appears to appreciate the words
addressed to it, and it responds to any act of kindness which
may be lavished upon it. Added to this, is the evident neat-
ness of its formation; the clearness of its various features;
the grace, as well as the lightness, of its construction, united
with speaking evidences of strength and of energy. The
quadruped appears fit to be the associate of man, and almost
seems upon an intellectual level with its master. As we contemplate the lustrous eye, and feel the rush of inquisitive
breath, it is impossible not to credit the tales narrated of the
creature's affection and of its generosity. We can, then,
sympathize with the love of the Arab for his steed, and sensibly feel that life in the desert would be rendered less deso-
late by the presence of such a companion.

Yet, this elegant quadruped is cast in no arbitrary mould.
Its beauty admits of the same variety which is conspicuous in
other animated bodies. The ears usually are small, and approximate towards their tips; but they may also be large, and
the points may be even wider apart than the roots of the
organs. Yet, in every shape, a thinness or a delicacy of the
outer walls, a nice arrangement of the internal protecting
hairs, together with a fineness in the investing coat, attest to
the purity of the parental stock.

A tribe of lop-eared thoroughbreds are known to exist upon
the English course: this peculiarity, however, is not a distinguishing mark of purity of blood, or a characteristic

running throughout the race. The fall of the ear exposes the interior of the organ to the eye of the spectator: that circumstance, no doubt, suggested the removal of the hairs, which Nature placed as guards before the opening. It is now a common practice, with almost every groom, to *singe* off these hairs with the flame of a candle. Such an agent cannot be safely entrusted to vulgar hands; probably, to this foolish custom is owing the deafness which by horses is so frequently exhibited. Any protruding hair the scissors might excise; but as regards the interior of the ear, grooms, had they even a slight acquaintance with physiology, would know that these hairs have their appointed uses.

GOOD AND ACTIVE EARS. LARGE EARS. EARS WIDE APART. LOW-BRED EARS.

With the ears no corporal excellence is connected, but with the health of this organ the general safety is associated; for the acuteness of the animal's hearing affords no mean protection to the rider. The absolute quietude of the ears indicates that sounds are powerless to excite the organ. Excessive restlessness of these parts suggests, that, by straining of one sense, the animal is endeavouring to recompense the obscurity of another: that the vision is either lost or imperfect. A lively carriage of the ears expresses a sprightly temper, and generally denotes a kind disposition: whereas, one member constantly directed forward and the other backward, is a frequent sign of "vice" or of timidity in its watchfulness.

Near the ear is the seat of another special sense. Many

people will pretend to discover the disposition of a horse by
the character of the eyes. A restlessness of the globe, the
display of any unusual quantity of white, and a perpetual
tension on the upper lid, are imagined to signify a " vicious "
inclination; but, in reality, these traits express only the
watchfulness of fear. Such indications are evidences of that
suffering which has been experienced; and these traits are
consonant with an anxiety to escape the future assaults of
brutality. Despair may not be desirable as a companion; but
it is not, therefore, to be falsely stigmatized.

A prominent eye, expressive of repose, and not exhibiting
an abundance of white, has been pronounced to be declarative
of honesty; though certain parties have condemned it, as

   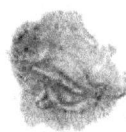

A WATCHFUL AND   AN HONEST EYE.   A LOW-BRED EYE.   A DISEASED OR
TIMID EYE.   PIG-EYE.

indicative of slothfulness. A quickness or activity, as contra-
distinguished from a restlessness in the visual organ, is, however, to be desired. The small eye, usual with the coarser
breed of animals, should be avoided, because, it is generally
accompanied by a heaviness of movement. The retracted or
deep set eye, which displays the organ only partially, which
is somewhat angular in figure, and which is commonly spoken
of as a " pig-eye," denotes either weakness of the part, or to
the majority of horsemen will suggest a previous attack of
specific ophthalmia. The disease, however, is not, in the
author's opinion, hereditary; but is generated by that close-
ness of abode, and that absence of ventilation, to which all
grooms strongly incline. The present writer has, most frequently, beheld ophthalmia in full and in perfect organs.

Before quitting the consideration of the face, it is imperative that the mouth and nostrils should be alluded to. In the well-bred horse these are both large, when compared with the same developments in the animal of a coarser origin. The lips should be smooth, soft, compressed, and suggestive of energy; but they should be without the smallest aspect of ill-temper. About them numerous isolated and long hairs may be located; but there should be no accumulation resembling a mustache, or bearing even a distant likeness to a beard. Such growths are commonly removed by the scissors of the groom; but the palm of the hand, if placed against the muzzle, is certain to ascertain the truth if those things ever have been in existence.

LARGE MOUTH AND NOSTRIL OF A WELL-BRED HORSE.   SMALL MOUTH AND NOSTRIL OF A LOW-BRED HORSE.   THE MOUTH AND NOSTRIL OF AN OLD, REJECTED, WELL-BRED HORSE.

The lowly-bred animal, being chiefly employed for slow uses, has not the need for those ample draughts of air which the faster speed necessitates should be rapidly respired; nor is the mouth declarative of the same determination which marks the lips of the purer blood. The bit is scarcely ever present upon the carter's harness, nor are the mouths of his charges formed to retain this invention. The characteristics of low birth cannot be effaced from the countenance of a quadruped. Age or privation cannot confound the two breeds. The thorough-bred in ruin is not to be mistaken for the teamster. No want, no suffering, no length of years, can obliterate the evidence of nobility from the animal of pure descent.

When purchasing a horse, it is always well to examine the angles of the lips. If any sign of induration is remarked, it signifies that the animal has suffered from abuse of the bit. If on any limited space, however small, a patch of white skin is observed located upon a dark ground, it denotes that " once upon a time " the true skin has been removed from that place, while cicatrix now exists to apprise future purchasers of the fact. If anything like a hardened lump should be felt in this situation, it demonstrates that the quadruped has a hard mouth, and is an obstinate puller, or that it has passed through the hands of an unfeeling master.

In either case, the creature is not a desirable possession. Harshness is not a kindly educator, nor does it beget docility of spirit in the being which is subjected to its exactions. A hard mouth necessitates one of the severest trials which can be inflicted on a horse proprietor. It is painful, every time a change of direction is desired, for the rider to tug at the reins; such a necessity soon destroys every pleasure of the exercise. But a regular puller is always a dangerous servant. Generally, it turns out to be a " bolter," and before running away will seize the bit between its teeth, when the driver or the rider alike is helpless. Our entreaty to the reader is to turn his back upon the offer, should he ever be solicited to buy a horse having a damaged mouth.

At this point, it is requisite the author should review the various organs which together constitute the head. An activity equally removed from stillness and from restlessness denotes health to be present in all the seats of special sense. These things are of more importance than at first glance is apparent; because such united testimony is the best security as regards the general system. It equally testifies to the soundness of the brain, and to the healthiness of the body. When the animal suffers, the perceptions mostly are inactive; when the brain is oppressed, the loss of sense first announces the disorder.

These organs also deserve attention for their own sakes. Man is not gifted with remarkable faculties either in seeing, in hearing, or in smelling. He therefore desires such assistance as the companion of his journey may afford. The value, consequently, of an animal is materially deteriorated by the loss of any of its protective powers. These, when all enjoyed in perfection, assist one another. When any organ is excited, the rest are seldom dormant. Thus, when the quadruped perceives in the distance some obscure object, the ears are advanced, and the nostrils are inflated. The same general movement is remarked whenever the hearing catches a distant sound, or whenever the scent detects a novel odour. All are conjoined to produce one result; therefore, the loss of one cannot be without effect upon the uniformity of action.

As regards the formation of the countenance, an enlarged cranium is no detriment; but the Roman nose sometimes interferes with the capacity of the nostril. When it produces such a result, the peculiarity warrants either a reduction of price, or an absolute rejection of the offered sale. In other respects, this make is regarded of no importance; but it certainly does not add to the appearance of the animal. Horses are generally prized in proportion to their beauty; nor can the author quarrel with such a foundation of judgment, as in most animals harmony of figure justifies a belief that excellence of spirit also exists.

The nostrils, however, are associated with the important function of respiration; therefore these organs demand consideration, when regarded apart from the other senses. They admit the air, which is inhaled by the expansion of the chest; consequently, the dimensions of the nostrils allow an inference to be drawn as to the capacity of the lungs. This opinion, however, should be only advanced after the alteration has been noted between their size when at rest, and their enlargement when excited. Should no marked variation be produced by the opposite states, then the value of the animal is only to be considered in connexion with slow work, as the speed must be regulated by the capability of receiving a quantity of vital air, proportioned to the power exerted.

After the capacity has been observed, the nature of the movements of the nasal openings should be noticed. Subsequently to exertion, ease of motion is not to be anticipated; but nothing approaching to spasmodic action should be remarked. The nostrils ought to be regularly expanded; not to fly open with a jerk, or to suddenly enlarge their form, as under the influence of a gasp. A capability of dilatation, attended with an evenness of motion, however fast the movement may be, are the points which should be looked for in the nostrils of a horse, because the characteristic changes attending inhalation best expose any defect in the respiratory apparatus; for by such a test the remotest disposition to become a roarer, or to exhibit diseased wind, is easy of detection.

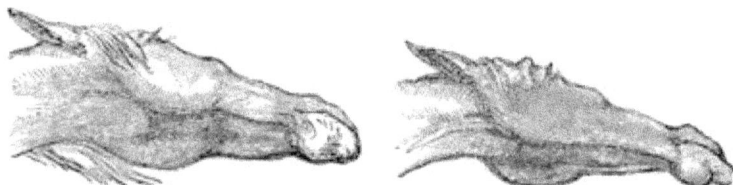

WIDE AND NARROW CHANNELS.

Connected with the head, every horseman comprehends how much width of channel, or of space between the branches of the lower jaw, is to be desired. The reason why such a form is highly prized in an animal of fleetness or of exertion is because such an opening allows room for the varied movements necessary for the offices of respiration or for the change of position imperative in the larynx, which is located near to or within the hollow thus provided. Clear space is, of course, imperative wherever rapidity of movement has to be executed.

There is also another thing equally desirable. That addition is a full development of the motor power which affects the larynx.

Wherever this last point has been noticed, the writer has always with confidence been able to pronounce the high character of the quadruped; he has not, in a single instance, been mistaken in his conclusion. The muscles, which are attached to the spur process of os hyoides, or to the bone which regulates the movement of the larynx, when well developed, are discernible in the living animal.

PROMINENT DEVELOPMENT OF THE HYOIDEAL MUSCLES.

They form a kind of indication, as though Nature was half disposed to invest the animal with a miniature dew-lap. They lead the muscles of the neck perceptibly more forward than these agents run in the majority of horses, and in some specimens they may, with a little manipulation, be traced almost to the point of their insertion.

The muscles last alluded to all originate from the trunk, the more forward cavity of which is known as the chest.

There is much dispute concerning the best form of the horse's thorax; but such a question can only be decided by the uses to which the animal is to be subservient. For instance, below is inserted the illustration of a cart-horse with an almost circular chest. Such a form permits the presence of a huge pair of lungs, and favours the increase of weight.

Sufficient oxygen is always present to convert the starch or the sugar of the food into fat : during slow work, enough of atmosphere to vitalize the blood must be inhaled, nor is excessive exertion calculated to materially increase the amount. Where weight is more desired than activity, where propulsion

A CART-HORSE.

is to be chiefly accomplished by bringing the heavy carcass to bear against the collar, such a make is admirable. All creatures, in which speed is not required, should possess circular chests; for by such a shape, the quadrupeds are adapted for the accumulation of fat, and for the performance of slow, of continuous, or of laborious work.

There are, however, numerous animals which are required

to possess capability for a "burst"; for the acme of which
phrase is embodied in the rush or the closing struggle of the
racecourse. The creature of speed, therefore, should exhibit
rather the deep, than the round thorax : for fat is not desired
on such an animal. The deep cavity, moreover, admits of an
expansibility, which is imperative, during the extremity of
muscular exertion. It is, however, sad to see well-bred

DIAGRAMS, ILLUSTRATING THE DIFFERENT CAPACITIES OF THE
OPPOSITE FORMATIONS OF THORAX.

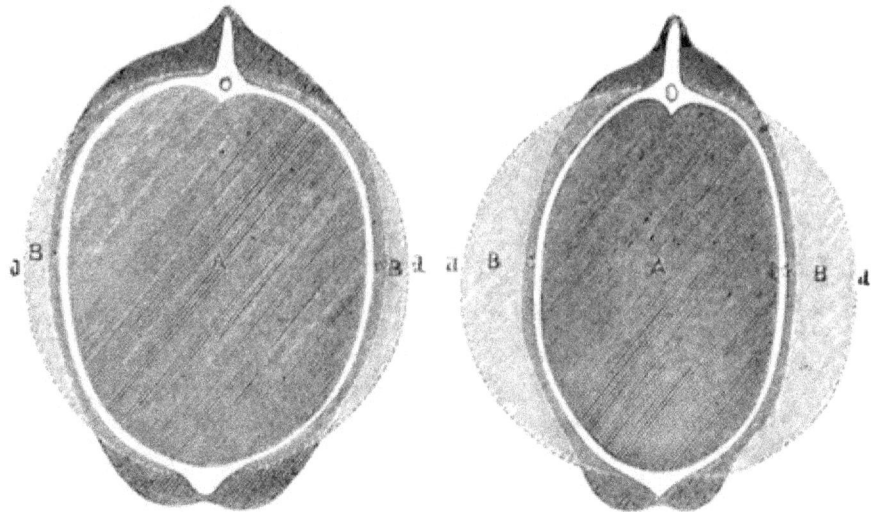

THE THORAX OF A CART-HOUSE.   THE THORAX OF A BLOOD HORSE.

A A. The capacities of the two chests in the quiet condition.
B B, B B. The limits of expansibility in each, when excited.
c c, c c. The outside of the coat in the quiet condition.
d d, d d. The surface of the body in the excited state.

animals in and about the metropolis forced to pull carts, for
which employment Nature has unfitted them. They possess
no weight of body with which to move the load. The burthen
must be propelled by the almost unaided power of the muscles.
The limbs, strained by the constant necessity of the posi-

19

tion, soon become crippled, while excessive labour causes the flesh to waste; hence, the miserable objects which are sometimes witnessed toiling along the thoroughfares of the metropolis.

To render the above facts comprehensible to the generality of readers, let it be granted that the lungs of the cart and of the blood horse, when expanded to the uttermost, would occupy the like space. When not excited, or both being of the normal size, the respiratory apparatus of the coarser breed is by far the larger of the two. In the passive condition, the heavy quadruped inhales much more oxygen than is needed to vivify the blood. The excess is, therefore, appropriated by the food and nourishes the frame: hence, dray-horses have a tendency to become fat. On the contrary, in the ordinary mood, the lungs of the thoroughbred receive scarcely more air than is required to uphold vitality; therefore, this kind of quadruped exhibits, as a general rule, no vast disposition towards excessive obesity.

During all quickened movements, however, the action of the lungs and the speed of the circulation are much increased. The impetus, given by motion to the vital fluid, causes the detention in the lungs to be of a comparatively brief duration. The period of change is shortened; at the same time a larger absorption of the vivifying agent becomes absolutely imperative. The greater depth of chest in the racer admits of a greater change of dimension: then air is inhaled, equal to the rapidity of movement. The pace, therefore, can be maintained with comparative ease. But the round form of thorax allows of little enlargement: the demands made by exertion cannot be complied with, and the heavy horse, when hurried, is consequently soon exhausted.

It is not, therefore, the size, or dimension of its thorax which fits the steed to the purposes of fleetness. That quality depends on the adaptability of the cavity to the exigencies of excitement; for such purposes, the quadruped with a round chest is not to be preferred. At present, there is no instrument by which the motions of the horse's ribs can be accurately ascertained: thus, the reader is forced to guess at an alteration, which cannot, under existing circumstances, be regarded with that confidence which is inspired by the knowledge of a fact. A quarter of an inch between the enlargement of the ribs in different animals (supposing the other points equal), should more than determine the winner of a race; since the change which takes place in the blood regulates the other properties of vitality.

The belly and the chest are distinct cavities; although there is communication between the organs of each. Thus the great artery, which originates at the heart, travels into the abdomen: while the veins, which traverse the larger division, also penetrate the thorax. Nevertheless, the contents and the uses of each space are generally distinct. The principal agents of the more forward cavity are the heart and the lungs; the thorax being chiefly sacred to the purposes of respiration and of circulation. The liver, the stomach, the spleen, and the intestines, are enclosed within the abdomen: the function of this region being engrossed by the offices of appropriation or by those of nutrition.

Most judges admire the horse which presents a belly apparently well filled by its contents. Certainly this appears to be the soundest of the many prejudices which appertain to horse-flesh. The shape of the thorax must, in no unimportant degree, regulate that of the abdomen: the two cavities being only parted by a fleshy screen, denominated the diaphragm. The herring-gutted quadruped is, commonly, as deficient in the respiratory as it is wanting in the nutritive functions. Of course, this rule is not absolute; but a capacious thorax is required to counteract any absence in the process of nutrition. The animal which rapidly narrows towards the flank generally

19 *

purges upon work, is commonly of a washy constitution, and usually possesses a bad appetite, frequently spoiling more fodder than it consumes.

HERRING-GUTTED HORSE.

Horses of the above conformation are soon found wanting in other respects. Narrowing towards the flank being accompanied with deficient quarters, enables them to slip through their body clothes, and renders it difficult to retain a saddle in its proper situation. The groom may in vain give extra attention to the fastenings, the dwindling form empowers little motion to displace the tightest of girths. The saddle always has an inclination to glide backward; and the rider, when such an occurrence happens, must be placed in no enviable position.

Objection even to a greater extent is engendered by the opposite kind of abdomen, or by one which is known as "a cow-belly," or "a pot-belly." Animals of this make always seem immatured; as though they had been brought into the world before the proper period, or had been forced to perform hard labour at too early an age; their legs are long, their

withers are low, their muscles are mean, their chests are
narrow, and their countenances are distorted by a querulous
expression. These unhappy creatures possess but little
strength for work; if made to travel fast, they are speedily
blown. In the stable, they are greedy; when out of it, they
are vicious. Many of their faults are to be attributed to
disease, the digestive functions being invariably disordered.
They are worthless, or are "all too feeble" for harness;
while the enlarged belly, when favoured by the motion of the
limbs, renders retention of a saddle an utter impossibility.

COW-BELLIED HORSE.

The legs of a horse,—these can hardly prove too short; for
brevity of limb is always an accompaniment to depth of chest
and proportionably powerful quarters. The long leg always
attests to the light carcass; hence, the motor agency of the
limbs is deficient, while the cavities of respiration and of nutrition are necessarily diminished. A narrow thorax almost
enforces low withers and an upright shoulder. The bone of
the arm, or the humerus, is pushed into an undue slant by the
forward position of the blade-bone or of the scapula. This
compels the front leg to stand too far under the body. Such
an arrangement favours neither beauty, speed, nor safety; in

fact, it is one of the worst forms which the components of the frame are capable of assuming.

The action of the shoulder-blade during progression is upward and backward, or it is drawn toward the highest processes of the withers. Low withers are, of course, opposed to extended motion in such a line. The lessened action of the bone necessarily limits the movement of the structures which depend from it, or the action of the humerus is governed by that of the shoulder-blade. The trivial motion permitted

A STRAIGHT SHOULDER, SHOWING THE POSITION OF THE BONES.

by low withers, therefore, limits the advance of the fore arm, the parts being, as it were, tied together. The natural carriage of such a malformation is with the head and neck protruded, so as to favour progression by strain upon the cervical muscles. At the same time the body inclines forward, which throws the limbs backward, or out of their proper situations; and this circumstance accounts for animals of this particular make so frequently encountering "accidents."

The gait, characteristic of an upright shoulder, is very peculiar. A bad fore hand is the most common defect witnessed in London thoroughfares. In the metropolis of the world it is, indeed, a rare sight to behold a carriage drawn by a pair of really good animals. The quadrupeds in general use for such purposes are mostly faulty about the shoulders. The fore hand is placed upon the trunk in too upright a position.

The job-master is conscious of this defect. He always endeavours to convince his patrons that such a make is advantageous where a creature is designed for harness. Possibly
the tradesman might succeed in persuading his customers
into a false belief, were not prejudice opposed to his sug-
gestions. Ladies admire high action in the steeds attached to
their vehicles; this is the kind of step which most of the
horses just described are incapable of long exhibiting.

Art or cruelty, however, can partially amend the faulty
motion of the limbs. Force the head into an unnatural attitude by the unscrupulous employment of the bit or of the

DIAGRAM, SHOWING THE NATURAL ACTION APPERTAINING TO A STRAIGHT SHOULDER, AND ALSO
ILLUSTRATING THE CHANGE SOMETIMES OCCASIONED BY THE UNSCRUPULOUS EMPLOYMENT
OF THE BIT OR OF THE BEARING-REIN.

bearing-rein; retain the neck erect, without regard to the
cramp induced, or heed of the strain cast upon the muscles,
and the torture, although the life be shortened, and the safety
of the owner endangered, nevertheless, may occasion the feet
to be raised during progression. This fact is illustrated in the

previous engraving. The natural mode of going is indicated
by the letters A A ; the possible change of form is to be seen
in the parts distinguished as B B, although the action there
depicted certainly displays a most unusual degree of amendment, to induce which must shorten the existence.

Any such improvement is always procured at a vast personal
risk ; for the head being raised partially throws the eyes out
of use. It also impedes the circulation, ruins the mouth, dis-
torts the body, and deranges the breathing. All these evils
are inflicted to obtain the kind of pace which is never natural,
but which closely resembles the sort of step that is characteristic of blindness in the horse. Few of the animals thus
treated live to descend very low in the scale of equine existence.
They mostly perish young ; but the reader may recognize them
drawing the broughams of gentility, and too often presenting
one of the cramped, forced, and uneasy paces which are depicted below ; for into such kinds of action all upon service
ultimately subside.

VERY FAULTY ACTIONS.

On the other hand, the animal with a deep chest and with
high withers, almost as a necessary adjunct, possesses a slanting
shoulder ; or, at all events, this probability is favoured by
that particular formation. Such an arrangement of parts
must be accompanied by an upright position of the humerus,
and the advanced location of the fore limb. This conforma-

tion is bettered materially by an arched crest, and a head
"well set on." Unfortunately these latter points are seldom
encountered; the proper disposition of the fore-quarter being
rarely attended with the last-named grace.

A SLANTING SHOULDER, SHOWING THE POSITION OF THE BONES.

Such horses, however, Stubbs, the animal painter, used to
delineate. Either the artist was particularly fortunate in his

A SLANTING SHOULDER IN ACTION.

models, or beauty has been sacrificed in the anxiety to breed
other properties. Such horses appear to have been common

in England, when the racer was compelled to possess endurance; and if report be true, the last animals exhibited a greater speed than their descendants can display. Hunters were formerly something better than the rejected of the course; they could show a beauty equal to their strength. Creatures with the fore hand such as has been described, are not only more pleasant to contemplate, but they are also capable of working with far less exhaustion to the system.

With a front limb of this nature, the movements of the leg are regulated by that of the shoulder. When the blade-bone is drawn upward, the humerus leaves its almost erect position, and assumes a forward inclination. This causes the arm to be advanced, and propels the leg and foot. Thus, the movement of a part governs the motion of the whole: a grace, or harmony of action is the result. The various components of the member change their relative positions to one another without effort, but with evident intention; all parts of the limb are simultaneously advanced. The work is not cast upon one set of muscles to the injury of another region. A well-made animal is one perfect whole, and formerly was common throughout the land. People may sigh that such quadrupeds are now lost to the nation: this regret, however, does not accord with the folly that upholds the racing mania, which has engulfed the once-prized native breed of English horses.

The articulated skeletons, which are exhibited in museums, present but poor resemblances of the living framework, as it is arranged by the hand of Nature. In these artificial preparations, the fore limbs are always straight, as are the supports of a kitchen table. But, contemplate the living example. The positive perpendicular is never observed. The member abounds in gracefully swelling prominences, and admirably poised inclines. The chest may be wide; but the hoofs are placed close together. Such a necessity renders an erect line an impossibility. Try the same rule in another direction. Let

a plummet be dropped from the point of the shoulder of a
living and well-made animal; it will mark the limit to which
the toe is extended, when the healthy horse is resting the
limb. Such a fact proves the sheer upright form of the
member to be an unnatural distortion, and a positive impossibility.

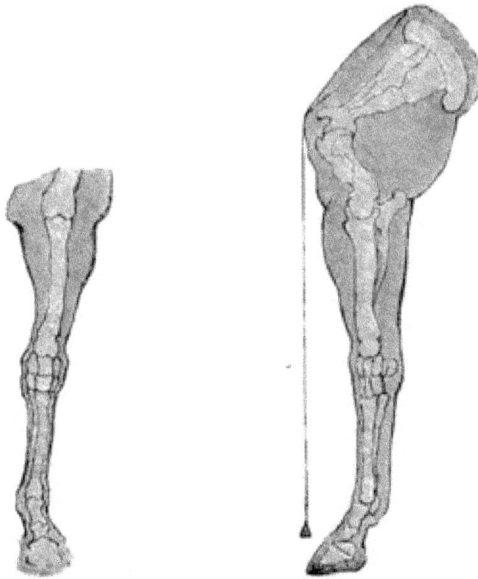

DIAGRAMS OF BONES WITHIN THE HORSE'S FORE LEG.

The importance of the shoulder and of the arm bone having
been enlarged upon, there remains to direct the reader's attention towards that which in general acceptance constitutes the
forearm, as well as the knee, the leg or the shin, the pastern,
and the foot. Where the limb quits the trunk, it should be
characterized by muscular developments; since, at this place,
resides the chief of that power, by which the lower portions
of the member are directed. The flesh should bulge forth,
and cannot be too abundant; for a thin fore arm is incompatible with goodness in a horse.

The point of the elbow should be prominently emphasized, as this bone affords a leverage, whence many influential muscles originate, and which some of the principal flexor agents directly operate upon. Towards the knee the swelling should gradually subside, leaving upon the surface of the joint a broad, clean, and firm appearance. At the back of the knee, there should stand forth, or rather should stick out, an osseous point, the size of which is of every value. Its aspect may not please the inexperienced fancies of the boy; but the uses of this development are, in no little degree, governed by its magnitude. It affords a point of insertion to the short flexors of the limb, as well as gives shelter to the perforans and perforatus tendons in their passage towards the pastern and the foot. Its magnitude, therefore, not only favours muscular action, but also indicates the dimension of those important structures which this bone protects.

The fore arm should be long; the shin ought to be comparatively short. The reach depends on the first, the length of which secures an extra amount of motor activity. No muscles of importance are located upon the shin: bone and tendon are the principal components of this region. The part should not be absolutely straight; for such a form is incompatible with all idea of living beauty; but, at the same time, it ought to present no obvious inequalities or sudden enlargements. The bone should be compact, giving to this portion of the limb, when viewed from the front, almost the appearance of being deficient in bulk; but when regarded from the side, the lower part of the leg cannot be too broad; for breadth and strength are here synonymous.

The above rule applies with equal stringency to both legs, —to the hind limb below the hock, as well as to the more forward member from the knee downwards. Each should be thin, when viewed from the front. Neither can well be too deep, when seen from the side. Both should appear solid, and each should feel almost of metallic hardness. The pastern joint should not present a level surface, when viewed laterally: and as it proceeds downward to join the foot, a graduated enlargement should exist.

Much comment is usually indulged upon the horse's pastern. The degree, in which this part may or may not slope, has been authoritatively defined. The reader will best judge of these

A LONG AND SLANTING PASTERN.

A NATURAL PASTERN.

AN UPRIGHT PASTERN.

AN OVERSHOT PASTERN.

INCLINATIONS OF THE PASTERNS.

opinions, by considering the purposes for which the pastern was created. Its intention is to endow the tread with elasticity. The fetlock of a racer, when the animal trots, may be seen to touch the earth, every time the weight rests upon the foot: nevertheless, the thoroughbred has, during the conten-

tion, to endure the very excess of action. There must, therefore, be something erroneous in the popular judgment which
connects weakness with the motion of this part, or no racer
could ever reach the goal; and if a quadruped does occasionally break down, the likelihood of such a misfortune is not
regulated, or to be foretold by the pliability of the pastern
joint. However, that the reader may estimate the value of
the prejudice, various pasterns, designated according to the
general phraseology have been submitted to his inspection.

To enable the purchaser to arrive at a sound decision, it is
necessary to state, that the inclination of this region is

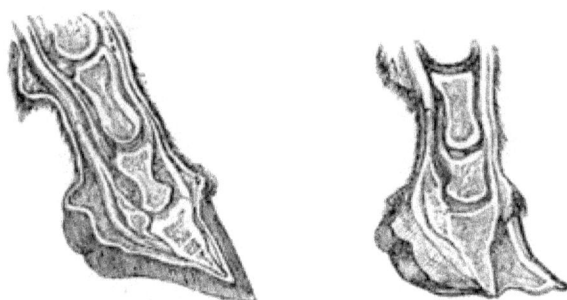

governed by the major flexor tendons, which are situated
underneath or behind them. Their slanting, therefore, is
regulated by no peculiarity in the forms of the bones themselves; but is controlled by, and dependent upon the condition of another structure. A short, upright pastern, if it
can bear any evidence at all, testifies to a stubborn and unyielding state of the great flexor muscles, the weight being
then thrown upon the osseous supports. The play of the pastern denotes nothing more than the healthy elasticity of the
flesh upon the tendon, proper to which, the osseous structures
repose. The bones have no motor power belonging to themselves. The upright and the overshot pastern suggest no
change in the more solid frame; but such alterations prove

that excessive work has strained the great flexors of the limb, and destroyed the inherent property of elasticity, with which every muscle is endowed by Nature. The burthen being then supported by an osseous pillar, instead of an elastic band, of course jar or concussion ensues upon the abnormal change.

Thus, alteration in the natural position of an oblique bone is of great importance to a purchaser, and to judge properly of the pastern joint, the substance, swelling forth beneath the elbow, must be regarded. Should this portion of the body be mean, or wanting in development, hard work will probably induce it to become rigid, or labour may, ultimately, cause the pastern joint to shoot forward and out of its proper situation.

The flexor tendon likewise influences another part. The perforans is inserted into the sole of the coffin-bone, or into the bone of the foot. The direction in which the toes point is, therefore, regulated by a substance, so far distant, that the attempt to connect the two organs may, to the uninformed mind, seem somewhat ridiculous. Yet, the statement being correct, the fact renders the position of the elbow of more importance; for, according to the situation of that bone, the hoofs will be directed. Thus, an ulna or an elbow, which is drawn towards the trunk, will be attended with a toe inclined outward. When the bone turns from the body, the forward portion of the hoof is directed inward. When the framework is properly constructed, the hoofs point forward; for horses' hoofs are liable to those derangements which the human foot exhibits, and generally with like results. Only, in man, striking one leg against the other during progression is not attended with the unfortunate consequences which such an occurrence often will induce when this accident occurs to the quadruped.

By the pasterns, recently exhibited, it will have been observed, that the inclination of the bones influences the slant of the hoof. The two structures are so connected one with the

other, that neither can be independent; for the direction of
the pastern, of course, determines the nature of the weight
imposed upon the foot. Thus, should the foot receive more
than a normal pressure, this circumstance, by throwing the
weight upon the bones, occasions the muscles to contract, and
produces upright or overshot fetlock joint. Nevertheless, the
hoof is operated upon by other agency. Diseased action will
also interfere with the growth of its outward covering. The
member may, under such injurious excitement, when long
continued, eventually become deformed.

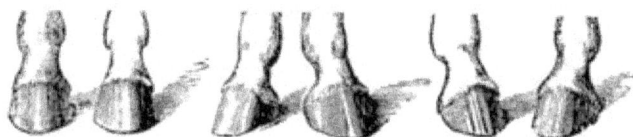

HOOFS POINTING FORWARD.   HOOFS POINTING OUTWARD.   HOOFS POINTING INWARD.

INCLINATIONS AND DEFECTS OF THE FEET, AS WELL AS SAMPLES OF ODD HOOFS.

The place of birth also influences the horn. Thus, a quadruped brought up on the fens of Lincolnshire generally displays a flat sole, a weak, a low, and a slanting crust. The
horse, whose native land is dry or sandy, mostly exhibits the
hoof high in the quarter and thick in its encasement. The
creature with feet of the intermediate sort, which a few years
ago were esteemed the model form, is generally the inhabitant
of a moist, but not of a wet district. The horn, therefore, is
indirect evidence of the rearing; and the author has now to
consider how far its condition can, by itself, be regarded as a
positive proof of any other fact.

There is one defect, not generally observed, but which
should always be studied in every examination of the feet. It
may surprise the reader, when the author declares it to be
very far from an uncommon circumstance to encounter a
horse with odd hoofs, or with feet of different sizes. Such a
peculiarity is totally independent of the defective inclination

of the toes, and may be seen in horn of any possible condition, or in feet of any variety of form.

An animal becomes lame in the foot. If the lameness is removed in reasonable time, the affection disappears, and leaves no trace behind it. But let it continue for months, and during such a period, the sufferer will throw little or no weight upon the diseased member. The part will be rested. The purpose or function of the organ will be counteracted by the will of the animal. The consequence of long disuse will be a proportionate decrease in size. Upon recovery, the loss of bulk is seldom restored; for if the foot is then employed, so also is the sound one; and the action being equal, of course, it does not particularly affect one extremity, but operates on both alike.

The difference in the feet may not be so startling as to enforce attention to the deformity. It is seldom of this nature. Most probably it will require some discrimination to detect it. In the last engraving, the author endeavoured to depict the defect as it was generally exhibited. None of the hoofs, there delineated, positively match: though, very probably the reader had not remarked their differences. However, the slightest disagreement is an accepted proof, that disease has been present,—at what time,—whether recently or long ago,—of what nature, whether structural or functional,—the examiner cannot tell: he however, assumes lameness has existed,—has endured for some period,—and he fears that the organ, which has been afflicted, may retain a liability to repeated visitations of a similar misfortune.

The so-called model foot is very liable to change, and not less likely to exhibit disease. It is very pretty to look at; but it does not, as a rule, undergo much work without alteration. This opinion, however, must be regarded only as announcing a general law; for though the intelligent Mr. Bracy Clarke puts forth engravings, illustrative of the effects which work produces upon the model foot, nevertheless, the writer of the present volume has seen hoofs of this description, which have, without apparent injury, endured constant shoeing, as well as perpetual battering upon the dreaded London pavement.

The slanting crust, weak heels, and low soles are, however, not to be commended. These are among the worst points which the equine form can present, and they are, too commonly, the forerunners of sad internal disease; as ossified cartilages, sand-crack, pumice-foot, &c.

After long reflection, the author must express a preference for the high or the stubborn hoof. When doing this, he is consciously opposing his unsupported opinion against the firmly and repeatedly expressed judgment of his professional brethren. He, therefore, can ask no man to agree with his decision; but he humbly requests the reader to peruse the grounds of his conviction, before hastily condemning its declaration.

The horse is a native of a dry and an arid soil. Such a region induces that which the inhabitants of this country stigmatize as an excess of horn or an abnormally high sole. This kind of hoof, therefore, would appear to be natural to the animal: at all events, such a foot must have been general before the invention of iron shoes. Moreover, when the immense weight of the creature's carcass is considered, and the manner in which bearing is increased by speed, is also properly regarded, a necessity for the stoutest hoof must be fully apparent.

In addition to the above inferences, the reviser may advance his own observation, carefully made through a number of years, that all animals exhibiting strong crusts are not, necessarily, cripples; but that the creature with such a development of horn is in consequence less, infinitely less, liable to pedal derangements. The contrary conclusion has been upheld, because most men thought the excess of horn must

check expansion, and also severely pinch the internal structures. With regard to the last deduction, all outward developments are produced by, and are governed by the inward
organs, which these shelter. The secreting member may be
soft, and the secreted substance may be hard : still, by a wise
provision of Nature, the tender structure rules the insensitive
material, which it produces. Therefore, the horn cannot press
upon or pinch the internal portions of the foot, any more than
the skull can compress the healthy brain, which it protects.

Then as to the supposed want of expansibility. The hoof
may appear stubborn, when between the human fingers; but
while supporting the body of a horse, it is exposed to the
operation of a force altogether greater than any which man is
able to exert. The question, therefore, is not whether the
hoof is very yielding; but whether it is so obdurate as to
resist the huge weight of the animal, when aiding the mechanical force of speed, and the vital action of muscular
power.

The author, however, while making the above declaration,
supposes form to be united with stoutness. Where the heels
have become " wired in," and the crust has assumed the upright figure, the internal structures must be in an altered condition, and the points of bearing for the different portions of
the limb must be entirely changed. The quarters in the last
kind of foot are, frequently, remarkably stubborn. They are
rather inclined to crack than to expand. Such parts will not,
by their innate elasticity, fly inward, on the leg being raised
from the ground, and thus regulate the amount of blood,
which shall be poured into the hoof: neither will they expand,
when the weight is cast upon the foot, and thus allow free
egress to the current, which is violently expelled in consequence of the superimposed burthen driving the fluid upward.

The upright hoof and narrow heels are, generally, all but
unyielding. They have lost their natural function, and the

harmony of the whole is destroyed. In consequence, the
blood, instead of being expelled from the hoof, cannot escape
from the pressure of the bony structures. The vessels, within
which the fluid circulates, are not formed to sustain uninjured
so vast a burthen. They rupture under the weight; hence,
this peculiar form of foot is commonly accompanied with
corns. Therefore, because corns are a disease, and because
disease, being once generated, is not in its course or duration
to be prognosticated with certainty, an upright hoof and
wired-in quarters are decided unsoundness; although, stoutness, simply considered, is rather a recommendation than a
defect.

The author may not dwell at greater length on this portion
of his subject; but those who desire further information, may
with advantage consult Miles's works upon the horse's foot;
which are the best, the cheapest, and the most lucid books
upon this topic in the English language. They are written in
a style which the most unlettered may comprehend; but when
recommending them, the author, in his own justification, may
state, that the views therein expressed frequently differ from
those opinions which are contained in the present volume.

Looking back upon such portions of the frame as have
formed the subject of the late remarks, there are certain
points which are invariably present in every well-made animal.
A very broad, full chest is advantageous for slow work; but
for slow work only. Where speed or activity is desired, depth
of thorax is indispensable; yet the cavity should not be
narrow, or the sides flat; while the exterior of the ribs should
apparently encircle sufficient space. The general contour
should, moreover, excite no idea of fixedness: the part should
convey a notion of its capability for easy and for rapid alteration of magnitude.

The abdomen should neither be large nor small. The exhibition of either failing announces a radical defect. The belly

ought rather to gracefully continue the line of the chest, than by its protuberance or the reverse, to enforce its existence specially upon the notice of the spectator. All may be considered right when the form elicits no remark; but when it challenges observation, the fact does not indicate that everything is as the purchaser could desire.

The position and the muscularity of the shoulder are the main points in the fore-hand. With respect to the limbs, these should leave the body as though they were parts of its substance. They can hardly be too large where they emerge from the trunk; and the fore-arm can scarcely be too long. The knee-joint should be broad and flat; while the bone which projects forth posteriorly should be well pronounced and evenly situated. The shin should be hard to the touch, and broad, when viewed laterally. The leg should seem straight and strong; the feet standing close together, and the toes pointing in a forward direction, rather than inclining to the outward or to the inward direction.

Such is a general view of organs, all of which are of equal importance. Breathing and digestion are such vital functions, it would be supererogation, did the author pretend to point out *their* importance. It may be otherwise with the fore limbs. Their use is not popularly comprehended; those members are exposed to numerous accidents and liable to many diseases. This predisposition is generally explained, by saying they are nearer to the heart than the hind legs are; and the straighter form is more favourable to a descent of the arterial current than is the angularity of the posterior extremities; therefore, this portion of the frame is more open to acute affections.

The facts stated are certainly correct. So is the less freedom allowed to the fore-legs, by confinement in, and fastening the head to a manger in a stall. Such, however, is not the whole truth. There are other causes in operation. The province of the fore limb is to uphold the trunk. Thus, at all times, the member has to support no inconsiderable burthen; but when that load is increased by the weight of a rider, or is augmented by the drag of the collar, the tug of the shafts, and the generally pendent position of the head, the reader may conjecture the force with which the limb must be driven to the earth, especially during any rapid increase of motion.

The continued battering to which the leg is subjected naturally exposes it to much suffering, which the comparative fixedness in the stable greatly aggravates. As the uses are severe, so are its afflictions painful; and it hazards nothing to assert, that very much of the sorrow which visits the animal is dependent upon the diseases or the accidents which are inseparable from these forward supports of the body and of the load.

When, however, the person called upon to exercise a judgment in the purchase of a nag is so new to the subject as to be incapable of forming an opinion, there is one primary test which seldom deceives; and upon the evidence thus evolved the merest tyro is fully qualified to pronounce. Let such a man mount the animal, and when seated in the saddle, he can surely decide whether he appears to be close upon the neck or placed far behind upon the back. A well-made animal, by the inclination of the shoulder and by the amplitude of the withers, forbids the forward location of its rider; whereas a worthless quadruped, by the lowness of the first dorsal spines and the upright position of the blade-bones, allows the rider almost to rest upon its neck; thereby, because of the greater weight to be supported by the front limbs, increasing the natural liability of the forward members to exhibit disease.

The reason why such a formation should be specially noticed, is, upon reflection, made apparent. The hind legs, by their greater motor power, always have a disposition to throw the weight upon the forward member. When this tendency is augmented by the burthen on the back, the consequence must be a destruction of any approach to an equilibrium.

The horse's body is, by Nature, given four props—one at each corner of the trunk. But when a human load is lodged almost over the fore limb; when the front leg is placed far behind the chest; and when the head swings in advance,—all approach to a proportionate amount of burthen is destroyed. The forward extremities then take a position almost in the middle of the substance, a proportionate encumbrance being removed from the posterior extremities. The hind members have less to do, and excessive duty is imposed upon the weaker organs, the motor machine being deprived of safety during progression.

While on the back, the rider should ascertain the shoulders are of equal bulk, or have not suffered injury; and that the trunk is sufficiently developed to secure a grip for the thighs of the rider.

The haunch is that portion of the frame upon which a capability for work is chiefly dependent. This region, therefore, should appear to be the embodiment of strength. It should not seem soft, or invite those pats which inexperienced horsemen are fond of administering to this portion of the body; but the aspect ought rather to suggest firmness and power; for here resides the force which must propel the load or direct the bound. Always choose an animal with good haunches, and invariably regard the position of the tail; as the situation of the dock, when on a line with the backbone, denotes the greatest possible length, and therefore the largest amount of muscular activity to be present.

Never purchase a horse which is recommended as fully equal to carry your weight; for the dealer who asserts this is, by his interest in the sale, incapacitated from forming a just opinion. But ostensibly appear to seek a horse for a friend— never for yourself—and state the nominal owner to ride at

least four, if not six, stone heavier than the would-be purchaser. There is a saying that an animal will run away with
too light a load; but that assertion is mere nonsense. Most
vicious quadrupeds are weakly creatures. The powerful frame
is generally united to an even temper. Strength does not
endanger the female equestrian; although ladies, generally,
are mounted upon the best-made, the strongest, and the most
valuable steeds. Indeed, this argument is never urged, save
when a gentleman hesitates to accept a weakly quadruped, or
desires to obtain the one which the dealer is not very anxious
should be purchased.

In illustration of this subject an engraving is inserted, which
represents better-made quarters than are commonly beheld on

A COARSE-BRED HAUNCH.

a native or coarsely-bred animal. But the reader can hardly
fail to remark that though the developments are not deficient
in width, yet the general aspect rather denotes softness than
expresses strength or suggests determination. The tail is well
set on for this kind of quadruped; still, the point of the rump
bone is not even indicated. The spectator must guess at its
precise location, as he cannot, by the unaided vision, detect
its exact situation. Bulk is not absent; yet, that which should

be its attendant is not prominent. The bones of the leg seem long, but the hocks are not remarkable for size or conspicuous for form. The limbs are not moved with that independence of action which gives to the step of the horse its air of resolution; but they are advanced, as though one was timid of proceeding too far without the other.

Yet, the inquirer may journey long and travel far before he will meet the equal of the quarters here depicted. The generality of these parts, on the animals of the coarser breed, are much narrower; the tail is seldom encountered springing from a position so near to the level of the spine; while, short as the extent of the posterior muscles may appear in the previous illustration, these are frequently to be seen of more circumscribed dimensions.

In contrast to the foregoing, the next engraving depicts the quarters of a blood horse. In this illustration, symmetry and

A THOROUGH-BRED HAUNCH.

beauty are equally preserved; but, with these qualities, also, are blended other attributes, which ennoble and elevate the object. Strength, power, and determination are impressed upon the image. Every muscle goes direct to the part on which it operates. The posterior line, on being traced from the dock to the leg, does not seem to hesitate between the bone of the member and the stifle-joint. The leg itself is thicker, but its greater substance depends upon the presence of muscle. The hock is cleaner, and uses of the part are better characterized. The os calcis, or point of the hock, as the backward projection is technically termed, stands forth prominently and affords the greater leverage for the motor agents to act upon.

When the quarters of the two breeds are contrasted, the difference is found to be extreme; the pervading attributes of each characterize the innate qualities of the animal to which the part belonged. The distinctions which divide the two are by these members well indicated. There was, some time back, a loud discussion as to what kind of horse was best fitted for ordinary purposes. The old staging days should have settled such a question; for, then, fast coaches found the employment of the nobler quadrupeds to their interest. Where slow and heavy propulsion is desired, the coarser animal is infinitely to be preferred. For all the gentler purposes of society, the thoroughbred is, in the author's opinion, to be recommended. Only, these fine creatures should be properly reared; they ought not, as now, to be produced with all the haste of greed, and cast upon the general public when found unsuited to the purposes for which they were generated.

It is offensive, if not painful, to hear persons speak of certain horses as though particular quadrupeds were created only for special uses. A good horse is fit for nearly every purpose; but such an animal is generally employed for the saddle. A thoroughbred, with lofty and well-developed quarters, is too valuable not to be appropriated by the racecourse. A blood, with so much bulk and stoutness as to indicate the qualities of endurance rather than of speed, is always destined to become a hunter. Horses of the purer breed are supposed not well suited for gentlemen's hacks. Good animals of this description are only too valuable for common purposes; but no creature is, by its intelligence, its activity, its gracefulness, or its beauty, so admirably qualified for the companionship of man as the noblest type of the equine race.

The manner in which the racer trots is asserted to express the action which is natural to all of the thorough breed. Before the reader agrees to that assertion, he should remember the trot is not a natural pace, nor one which the racer is broken to exhibit. Seen upon the course, the foot evidently moves too near the earth to clear the ruts of most English highways. Yet, as there shown, the motion is not to be despised. During it, at each step, the limbs are extended; the reach is admirable, and affords a far better foundation for excellence in a hack than the high action which is so highly esteemed by many. Great speed without great exertion is secured by the thoroughbred trot: much ground is covered, whilst the pace is easy, and pleasant to the rider. It is very opposite to that which medical gentlemen occasionally recommend as a " hard-trotting horse." A child might sit upon the back of a well-bred steed. The author recollects to have only seen one animal of this description employed as the riding companion of a gentleman. The master (a northern nobleman) was evidently proud of his possession; for the hack abounded in energy and with fire. The life never appeared fresher in a colt; but, on inspecting the teeth, the writer was pleasantly surprised to behold indications which denoted that, at least, twelve years had been passed. The foregoing illustration will suggest to the inexperienced reader the more striking peculiarities which characterize the well-bred action.

Any quadruped is supposed suited for the collar when it displays points which unfit it for the saddle. A prime saddle-horse, however, always makes the best harness animal; only, it is considered too valuable for such a purpose. There is but one law, which is absolute with draught horses. In them, the fore legs are pardoned a few faults; but the hind quarters

should always be powerful. That is desirable in all quadrupeds; for draught of every kind it is essential : it should never
be overlooked, or the want of such a property ever be pardoned.

There is another point of importance. Any gentleman
purchasing a draught horse—no matter whether for cart, for
carriage, or for phaeton—be it for any kind of vehicle, he
should be certain, before the transfer is concluded, that the
new possession stands high enough. Nothing looks worse
than small horses before a tall carriage. The living power
may be in excess—it can hardly appear too mighty—but an
inch below the requisite size gives to the most elegant and the
newest of " turn outs " a shabby and a mean appearance. The
draught may be light ; the horses may not be overweighted ;
—still, no fact or knowledge can reconcile the eye to the
general effect, where animals are too small for the machine to
which they are harnessed.

Of recent years there has been displayed a desire to infuse
the Eastern blood into the heavier breed which is native to
this country. The desire was commendable ; but its gratification has led only to evil. It has enabled the dregs of the
racecourse to be palmed off upon the public. A thin and
lanky offshoot of thoroughbred stock can be of no value.
These things should not be bought by gentlemen for any kind
of service. The time has come, when it is simple prudence
that the public should refuse longer to absorb the cast-offs of
the stud farm. No doubt, before the breeding of blood stock
became a general practice, the infusion of Eastern fire and
activity was a national boon ; for, a reference to engravings of
a few years back exhibits the animal, suited only for a plough,
used as ladies' palfries. The following copy from a figure,
presented in the famous folio work by a former Duke of Newcastle, will give the reader some notion of the kind of horse
once chosen to carry the fairest portion of creation in the
British isles.

From the following illustration, which may be well supposed
to embody the height of fashion and the cream of style shortly
after the accession of Charles the Second, the reader can
imagine the practical knowledge possessed by those writers
who speak of James the First as having greatly improved the
native breed of horses, and quote the benefits conferred
upon the national race by the more temperate but equally
determined enthusiasm of Cromwell operating in the same
direction.

LADY, HAWKING.

At this place, the reader must have patience while the method
of judging the limbs is pointed out. When the dealer exhibits
an animal, the customer's eye always should endeavour to
ascertain the bulk or substance of the creature which he is
expected to purchase. To do this, let the eye be directed

towards the chest, to ascertain if the fore legs are separated
by any breadth of thorax, or whether they spring from the
body almost from the same point. This decided, a glance
may be given to the line of the fore legs; these parts, also,
can be viewed as the gentleman passes backward. Having
reached the last situation, he observes if the thighs are large
and fleshy, keeping the legs well asunder; also, whether the
hocks are rightly placed, are huge, and are cleanly shaped.

Such remarks are important, since the disposition to cut is
generally decided by the width of the horse's trunk. Any
deficiency in this respect indicates weakness, as well as declares a general unfitness for severe labour. This circumstance being observed, it is usual for the horse to be run up
and down the ride. While the limbs are in motion, the spectator should notice the peculiarities of their carriage. A
flexion of the front shin to the outer side warrants a belief in
the existence of a splint. When the hind limb is not properly
flexed, but the toe is allowed to graze the ground, it is a
positive proof that the hock is disabled by the presence of a
spavin.

A worse evil, however, is, when the fore legs, during progression, crossing each other, the trot becomes a sort of
" hand over hand " pace. This kind of action is accompanied
by "speedy cut," or by a wound made upon one leg, immediately below the knee, with the shoe on the opposite foot.
That defect justifies an instant rejection; for such a liability
is incompatible with safety, as the blow too often brings the
animal and its rider to the earth. The legs being close is the
cause of "brushing" or of "cutting,"—a most troublesome
defect, which inflicts a wound considerably nearer to the
ground than "speedy cut."

Before purchase, the hair on the inner side of the legs
should be carefully examined. If a cicatrix or a bare spot is
discovered near to the seat of cutting; if any paint or colouring

matter can be detected on the part; or if the hair does not lie
perfectly smooth upon the place of injury, have nothing to do
with the animal. It is quite true that most fresh and nearly
all young horses will cut—others strike only towards the end
of a long journey; but it is also true that particular horses,
however fresh, or however tired they may be, never strike or

AN EXAGGERATED VIEW OF A WEAK ANIMAL, WITH DANGEROUS ACTION.

cut. The quadruped which a gentleman desires is one that
does not contain evidences of a liability to accident or to
disease. He wishes for a sound animal; and one disposed to
strike cannot be so esteemed. Every man wants a horse for
service; but the creature which may at any moment receive a

wound that shall incapacitate, assuredly cannot be esteemed a
serviceable possession, in any meaning of the word.

While examining the legs, the gentleman should also notice
the shoes upon the different feet.   If these are rusty, the fact
demonstrates that the horse has been wearing wet swabs, and
has been long stationary in the stable.   The circumstance is
suspicious.   In horse-dealing a justifiable suspicion is always
acted upon as an established fact.   If the shoes are of rude
make and much worn, it looks badly; and though it is no
recommendation, it justifies no inference.   But if the shoes
be thicker at one part than at another; if the horse, being a
nag, should wear very high calkins; if the toe be shortened,
or one side of the metal is obviously narrowed, it denotes precautions against clicking and against over-reaching; the first
being a most audible annoyance, which may lead to the forcible
tearing away of a fore shoe, and the last causing a fearful, a
terrible, and an incapacitating wound upon the heel of the
fore leg.   Also, should the toe of the hind shoe be ground
down, while the heel exhibits no obvious wear, the fact demonstrates the existence of a spavin.   The reader is therefore
advised not to purchase a horse on which any of the above-
enumerated defects are noticed, as such defects constantly
cause unsoundness.

CHAPTER XII.

BREEDING.—ITS INCONSISTENCIES AND DISAPPOINTMENTS.

THE greatest mistake that a man about to commence breeding horses can make is to select for such purpose old
mares. This one fault has produced weak and badly-formed
animals, who in their turn, perhaps, have propagated worse
offspring. The bodies of most old mares have been weakened
by hard work, or crippled by too early labour: consequently
the foals of such creatures are before birth injuriously impressed
either in shape or constitution; and for this reason a breeding
mare, by the majority of farmers, is considered a bad speculation. The same remark relative to early labour applies to the
growers of blood stock; both sire and dam are, before the
second year of babyhood has been completed, placed in the
trainer's hands in order to undergo preparation for the racecourse. Certainly, where an amusement is pursued with a
reckless defiance of economy, a little longer grace might be
accorded to the animals employed to promote it; or where the
topmost prize is estimated, not by tens but by thousands, it
might be prudent to speculate with a little forbearance for
such a reward.

Has it never occurred to a nobleman, or to any gentleman,
that it might, probably, be as profitable to keep the most
promising foals sacred to breeding purposes; that, simply as
a paying speculation, it might answer to do for the course
what agriculturists have done for the land, only with this

21

difference,—that whereas one desires bulk, the other should
aim at courage, strength, and speed. Animals, if well cared
for, and never placed in the trainer's hands, would in all probability bring forth finer specimens of horseflesh than either
their parents or their progenitors. These foals, being selected
and kept apart until the sixth year, might generate young
which would sweep the land; and a stud of " clippers " would,
assuredly, prove a pretty private property.

How far a youth passed in running improves the reputation
of some quadrupeds is well known; but how far it is a good
preparation for the offices of paternity is exemplified by most
blood mares and stallions becoming famous through their
progeny only after years of repose have mitigated the chronic
evils of their early life. Prejudice, however, takes no heed of
such teaching, but maintains the absolute necessity of proving
both before sire or dam are allowed to perpetuate their kind.
The consequence of this system is shown in the deformed and
misshapen dwarfs which are now ruining the once-prized native
breed of English horses.

What are results of such a system? Distances have to be
shortened. Many start; but few return, contesting the race.
Ages have to be altered; while boys have to assume the cap
and the whip. Useless weights are sought to suit the failing
strength; but more animals break down in the training than
come to the post.

Yet, racing is maintained, not for the amusement of a few,
but *to improve the national breed of horses !* How far does it
answer its purposes? Let the public markets testify. A stout
hack is a rarity. Such an animal was once all but universal.
A brougham horse—one looking fit to pull a house—was formerly to be found in every yard. Now, London shall be
searched through before the shadow of the original can be
encountered; when discovered, the price demanded will be
far too heavy for the generality of purchasers. The horse-

flesh of England is becoming weedy under a forced system. Poor "bloods" are everywhere present. In the sphere to which this breed should be confined, a few foals are retained; but the majority are discarded. Many are born that do not return the first expense which called them into existence. Those rejected are to be seen drawing cabs, carrying riders, pulling carts, and performing every office, which is at once a proof of their utter want of value and the hollowness of the pretence which perpetuates such degradation.

The glut of worthless "bloods" serves to check the raising of the other and the better kind of animals. The refuse of the stud farm being disposed of to the highest bidder, so far keeps down the price of common horses, that, what are termed serviceable quadrupeds have become scarce throughout the land, which once produced them in abundance. Thus blood stock is contaminating the native breed of the country. Even with particular breeds—as with the Cleveland bays—the horses which dragged the cumbrous vehicles of our ancestors are lost to the present generation. Carriages are built lighter; but the animals, being nearly pure blood, lack strength and want substance.

All animals, which are intended to perpetuate their race, should be comparatively young, and only subjected to such easy toil as will repay the difference between the stable and the field. The quadrupeds should be daily groomed, and ought to be supported by fodder of an extra nutritious character. Gentle labour and a warm loose box will only keep the body in good health. When not required to work, the animal should be left at liberty to roam about a piece of bare pasture; especially during the night, when the flies are not abroad, and when the vision of the horse enables it to move with perfect safety.

This treatment should be continued almost to the time of foaling: when the period is very near—three weeks or a

21 *

month of perfect rest may be accorded; duration being regulated by the condition of the animal. Rest, however, does
not imply that the expected mother is to be turned into a
straw-yard, or is to be exposed to the inclemency of the season.
One month subsequent to birth, the work may be gradually
resumed; but the mare and her foal should not yet be made
to travel on the high roads. The little life may, in the fields,
safely gambol by its parent's side. The exercise will benefit

BLOOD MARE AND FOAL.

the youngster, while its eye will become accustomed to the
toil with which it will have to be associated hereafter. But
the tender hoof of the newly-born is not, at the expiration of
the fourth week, so formed or so hardened as to endure the
grate of the common highways, although the feet may sustain
the wear consequent upon moving over meadow land.

The foal, before it saw the light, would be sustained by the

good food consumed by its mother; the mare would not, by
gentle work, be so lowered, as to unfit the quadruped for the
offices of maternity. By selecting the jobs to be executed,
these need not require greater exertion than would be necessitated by healthful exercise. Thus, a suggestion, which
to many minds may appear a heartless exaction, being explained, becomes no more than a conservative recreation.
Something of the kind is needed, because gestation and lac- .
tation naturally dispose to sloth, and half the danger of par-
turition springs from the debility which idleness engenders.

To render this subject more easily understood, let the
reader ask the family medical attendant, who is blest with
the strongest child,—the wealthy lady, who can afford to
repose throughout the day upon a sofa, or the tradesman's
wife, who is necessitated to bustle about, and to assist in the
lighter portions of the household duties? Or, if a more direct
illustration be needed, it is afforded by the contrast presented
between the swarming cabin of an Irish labourer and the
often heirless mansion of the English aristocrat.

Were such a custom, as has been indicated, only prevalent,
those "stud-farms," where mares are taken in and confined
in the straw-yard, with newly-born foals by their sides, would
be thrown out of use. The animal, being daily harnessed,
would be constantly inspected. There is always plenty of
light employment for one horse, if a farm be kept in order.
These odd jobs are now either neglected altogether, or are
suffered to accumulate, until a waggon-load of rubbish en-
cumbers the soil. To remove such heaps and obstructions
from time to time, the mare and a boy might be profitably
engaged, doing quite work enough to pay for corn and to
recompense for grooming. The necessary handling would
prevent that condition of semi-wildness, into which too many
mares degenerate; while the nature of the labour would not
render it profitable for a proprietor of land to keep more than

one quadruped for breeding, which is the number that most
farmers could find leisure to attend to without neglecting
other things.

THE BROOD MARE.

During the months of gestation, the animal should be
fondled and caressed. Any kindness which may be now
lavished upon the submissive slave will be certainly repaid
hereafter. The hour is approaching when a familiarity with
man may soften restraint, and render less perilous the time
of danger. The mare being more intelligent than the cow,
feels more acutely, and does not suffer so apathetically. It is
more demonstrative in its behaviour; but the generous quadruped will, in the utmost wrench of agony, recognize the step
or the voice of one who has been kind, and will even be sustained by the presence of him who has earned its confidence.
The animal is by gentleness wooed, as it were, to submission.
It learns to associate happiness with the person of its superior,
and willingly subjects itself to his assistance.

As before suggested, let the most promising filly be destined
from its birth for the function of breeding : she should never
be placed in the hands of a "breaker," or have her back
strained by being mounted. The creature should be rather
coaxed to toil, than coerced to labour; it is astonishing how
much more can be accomplished by such means, than will be
effected by the harsher methods. Subsequently to the fourth
year, the quadruped may earn its keep; but it should never
be urged beyond that point, and where a difference must exist,
the balance should stand in the animal's favour. Only the
lightest jobs should be chosen,—the mare being treated more
like a favourite slave, than regarded as the servile drudge,
whose exhaustion will tend to the profit of a harsh proprietor.

In this manner the first six years should be passed, when
the mare, being matured sufficiently, and uninjured by work,

may be put to the destined purpose; similarity—not sameness, but more decidedly not difference—regulating the choice of a sire. In the selection, allow one to amend the faults of the other; but in seeking this, avoid absolute contrast, as the union of opposites is too apt to produce deformity.

When choosing a mare for breeding, endeavour to discard much which has been printed on this subject. Let compactness of form, strength, and an aptitude for exertion decide the choice. The legs should be stout and short,—declaring bone and tendon to be present. The upper portions of these members cannot be too bulging, thick, long, or muscular. The crest should be highly arched, and characterized by substance; for the movements of the body are much controlled by the muscles of the neck. The shoulder cannot be too fleshy, so it shall slant properly, is firm to the touch, and is situated below withers sufficiently lofty. For hunting or for ordinary purposes, high withers are imperative. For racing they are no recommendation, as lofty action delays speed, and lessens the length of stride. The back should be short, save only in the racer. The loins ought to be broad. The hips cannot appear too ragged or be too wide apart, while the quarters must seem large in every direction; nor is it to be considered a fault, should these last parts stand higher than, and appear disproportioned to, the other regions. Above all, see that the channel is wide, the mouth large, and the nostrils ample.

Do not, according to the prevailing notion, search after a long or roomy trunk. Most people like such a shape because the carcass which they seek after is wanted to contain, with a foal, the enormous quantity of grass which the animal is forced to consume before life can be sustained. The mare just described is not supposed to live in the field, but to be as carefully tended and as liberally nurtured as the best horse in the stable. It is, during gestation, desirable that nourishment should occupy as little compass as possible, while it should not corrupt the body's natural juices. This last effect is consequent on the consumption of dry fodder. The moisture of the mother's body is abstracted from the fœtus, to soften the harsh and hard food which oppresses the stomach. But when grass is eaten, an excess of water renders that which should support the growth of the future foal weak and devoid of nurture, while it engenders dropsy in the dam, and also compresses the dawning life in its primary home.

When the period arrives, the time occupied by the mare in "foaling" will be short. The cow is usually slow in these matters. The mare is always speedy, and far less patient under pain. Therefore when the signs, which are well understood, declare the time to be rapidly approaching, send immediately for the nearest veterinary surgeon. However, previously ascertain that he is apt in this kind of business; and, above all things, be sure he is a feeling man. A coarse and noisy practitioner is of no service about horses. The words may not be understood, but the manners are quickly interpreted. The quadruped at this period wants support, encouragement, and kindness. A harsh command or a threatening gesture may so alarm timidity in its hour of excitement as shall retard the event they are intended to facilitate. Severity, however, does not always lead to any immediate result; but it may so flutter the mother or disturb its system as will assuredly be fruitful in after-disorder.

Should the animal be properly formed, and have been well selected, but little aid will probably be required; yet it is always prudent to have assistance at hand, as the mare on such occasions admits of no delay. Do not, however, allow the animal to give birth in a field or in the open air. Such may be the prevailing custom; but custom is always a bad leader for a prudent man to follow. Numerous children are born under hedges or in gipsy tents; but, nevertheless, such places are not to be preferred for ladies; and the horse now under consideration has not been reared upon a common, nor is it one that knows only comfort during the presence of sunshine. Lead the quadruped gently into a thickly-littered, loose box, having trusses of straw carefully poised against the inner walls of the building.

PREPARING THE LYING-IN CHAMBER.

The proprietor, however, must not be regulated in his measures by any rigid attention to dates. These afford nothing like an absolute rule, worthy of being implicitly obeyed. Neither need he be thrown into a fluster because the mare heaves at the flanks. Such a symptom, when unaccompanied by other signs, merely denotes a passing spasm, which may generally be removed by the following drink. Should the pain not yield, the dose may be repeated in half an hour; for, at this critical period, no bodily disturbance can be without importance. These attacks are said to be produced by drinking largely of cold water, by unexpected excitement, &c. &c.

Drink for heaving of the flanks.

Sulphuric ether   .   .   .   . One ounce.

Cold water   .   .   .   .   . Three-quarters of a pint.

Mix : Stir till the ingredients are blended ; then give as gently as possible.

THE NEWLY-BORN FOAL.

When the foal is born, let it be received in the arms of the groom, and with care laid upon the straw. This done, all present had better retire, for the mother and its offspring may with confidence be left to Nature. There should be no peeping through crevices, for the eye of maternity is cunning at detection. Neither should the slightest noise be permitted around or near to the building, as the nerves are always morbidly excited during this particular period. Silence is a good medicine to quiet a disturbed system. The creature will do well if left to itself. The cleansing of the foal may be confidently trusted to the parent's affection. All she immediately requires is a pail of milk-warm gruel; three hours afterwards she may accept a meal of prepared food.

Should the after-birth not be immediately ejected, resort to no purging; neither adopt any mechanical contrivance to expedite its expulsion. These old methods are altogether wrong. The retention is caused by the weakly condition of the mare, which allows the uterus to remain relaxed. The fittest physic in such a case is a quart of strong and sound ale. Give three doses of this medicine, each administered after a lapse of three hours. Should no effect have resulted subsequently to another pause of the like duration, inject into the part a full stream of cold water, permitting the fluid to return unchecked. Continue to do this till a spasm appears; then leave off, for your object is accomplished: the pain announced the viscus has contracted.

Dry the mare; give another pail of gruel; place a feed of softened food in the manger, and leave the creature to luxuriate in that rest which will now be enjoyed.

Animals soon get over such affairs. The foal requires nothing beyond a sheltered abode and its mother's attention. Should, however, the source of the young one's nourishment prove unprolific for more than twenty-four hours, a little skimmed cow's milk, first boiled and then slightly sweetened, being afterwards diluted with its amount of warm water, may, if sufficiently cool, be presented. The human hand is inserted into the fluid, and two fingers only allowed to protrude above the surface; these are generally seized upon, the nourishment being easily imbibed by the hungry foal. More than a single feed is seldom needed; even that had better be withheld, until evident weakness necessitates its administration.

Do not bother the mare or be tempted to thwart the course of nature at such a time with the impertinence of ball or drink. All physic should be withheld. The common Parent is very indulgent at such seasons; unless opposed by mortal ignoranee, her kindness generally proves the best restorative. However, should the bowels continue decidedly costive, some abdominal irregularity may be suspected, and then a bran mash, into which some softened corn should be thrown, will commonly afford all requisite relief. With regard to the newly-born, it is better not to interfere. So the parent be kept in health, the offspring, usually, has all the medicine it requires. Liberal, not too stimulating diet,—a sheltered abode, a dry ground, and a kind proprietor, embrace the chief, if not all, the wants of an animal in this condition.

The mother, after her title is confirmed, should always receive her food out of some vessel, which a man should hold, during the time it is consumed. Much good is thus effected by allaying the fear natural to maternity; the person, so occupied, should carefully abstain from any act which might alarm the anxiety of a newly-made parent. The same individual should not always present the meal, but different people should assume this office: so the animal may be thus trained to regard men as friends, and taught to depend upon the generosity of its superiors. By degrees, the foal should be coaxed to accept morsels from the hand of its attendant; advantage should then be taken to pat and to fondle the timid youngster. The purport of such lessons is quickly understood; for the horse appears, naturally, to value any act of condescension for its master.

It is usual to re-introduce the male a few days, generally three, subsequently to delivery. But such a custom is far too saving to be profitable. How does man imagine that one poor body is, besides extracting sustenance from grass, to yield milk to the living and to sustain the growth of the future offspring at the same time? It has been well declared, that no organ is equally fitted to perform two offices; but surely either of the functions alluded to is a sufficient drag. Moreover, to demonstrate how these functions are opposed, a fact of common occurrence, among the lower order, may be mentioned. When failing wages render an increase of family undesirable, it is usual for the married women to suckle the last child, even for years; thereby delaying the advent of the next intruder.

To afford the nutriment which shall maintain two growing lives and to support itself, is obviously too great a tax to be readily sustained by one body. The drain must be the greater because each will demand the more as time progresses; thus,

BREEDING, SUCKLING, AND LIVING ON GRASS.

the unborn has a portion of its sustenance diverted, while the milk on which the living foal should be matured is impoverished by the necessities of the maternal system.

Therefore, when entering upon the speculation of breeding horses, it should be remembered, that though a foal is a foal, nevertheless, a good and a bad foal are very different beings when tested by figures in an account-book. One good foal

every two years will pay far better than four bad foals every
year, as the eight indifferent creatures may be well sold at
£20; whereas a promising produce may be purchased at a
very reasonable price, if it should be parted with for no more
than £50.

There is a silly method commonly adopted by horse-
breeders, viz., that they, whether the foal be healthy or
weakly, only permit it to run by its mother's side for an
arbitrary period. Should the young one be strong and well
developed, and six months old, then, and then only, can the
maternal attentions be dispensed with, as, often, the too early

THE OULD MARE.

weaning retards the growth of the youngster, and, in so doing,
checks for ever perfect development.

There is another absurd custom in which the farmer delights. Having hacked and hunted a mare until her nervous

system is impaired, and her body worn, he reflects thus :—
"That ould mare has proved a downright good bit of stuff.
I should like to have a foal out of her before she is knocked
on the head." So he procures the service of some led horse,
and turns the aged animal out to grass, "to rest herself,"
as he asserts. This conduct would seem to be the climax
of possible folly! Nevertheless, the farmer acknowledges
nothing wrong in his behaviour; for he is as bold and as loud
in his lamentations as other people, when a weakly foal results
from his want of consideration.

CHAPTER XIII.

BREAKING AND TRAINING.—THEIR ERRORS AND THEIR RESULTS.

IN language, manners, costume, or in any of the many things which mark a *people's* advance, fixedness has not been allowed to check invention; but, where improvement was most needed, not only to ameliorate the condition of the slave, but to confirm the progression of man, by rendering impossible those sights which degrade and which debase the reasoning faculty, it has, apparently, been absent. The creature, during these years, has altered in form, and has become milder in character. The spurs and bits of former times are no longer in general use; because these are no longer required. They, assuredly, were not cast aside from any consideration for the life, to coerce which they were employed; although, a simple regard for property may have banished such ready instruments of torture and of injury.

The lunging of the existing horse-breaker is obviously nothing beyond that circular practice which constituted the chief portion of equine education with our forefathers. In an old book entitled *A General System of Horsemanship*, by a late Duke of Newcastle, it is depicted over and over again, until the image, from repetition, grows tedious. It seems very difficult to understand the useful or rational purpose which this peculiar lesson is now intended to support. Some persons assert, it is of much service in taming, as it assuredly must tire the colt. Others declare, it teaches the animal to bear properly on particular limbs. A third party assures us it is of infinite service, because it instructs the young horse in leaning towards the rein, and, by not permitting the eyes to be wholly engaged in directing the feet, it obliges the quadruped to employ "high action."

The use of the limbs is governed by the natural formation of the body; this last no Breaker will undertake to improve. It, certainly, is assuming too much for any art, to pretend it can alter that which Nature has decreed. A well-formed creature, although it should never have experienced the Breaker's instruction, will, of necessity, exhibit grace in its movements. The action of a badly-made quadruped may be temporarily disguised, but it will permanently retain only the mode of progression it is fitted to exemplify. By forcing a faulty horse to trot in a shallow stream, or by obliging the animal to move briskly with sand-bags attached round the fore fetlocks, a badly-made colt often will, for a space, adopt a higher action; but it is always certain, that this step, which has been acquired at personal inconvenience, will not be long maintained when the inducement no longer operates.

But, to take a practical view of the good likely to result from lunging:—Horses sometimes are obliged to move in circles. Mill-horses pass their lives in such educational employment; the only effect produced by this long course of instruction is that the poor victims become sightless. Travelling round and round soon causes giddiness, or induces a determination of blood to the brain. Young animals often stagger when relieved from their monotonous course of lunging duties. Old quadrupeds, we are told, grow used to the motion; but such familiarity is purchased with the deprivation of one "precious sense." This termination is hastened with the rapidity of the movements. Mill-horses walk their monotonous rounds; but the Breaker, dreading no results, makes the colt trot when describing this, his favourite figure.

22

Blood, therefore, rapidly loads and oppresses the brain of the young animal thus abused; and this consequence is the quicker as the pace is more excited, because the circulation is not only faster, but it is also more under subjection to external influences in the young than in the matured. The optic nerves originate from the sensorium, being a direct continuation of the substance of the brain itself: whenever the nervous centre is congested, sight is the first sense that suffers, or the first that tells the condition of the organ. Frequent repetition of this result upon the delicate structures of growing

LUNGING.

life appears to be an antiquated custom, which modern civilization should immediately abolish. It is not prudent in man to hazard the injury of his most valuable possession, when he simply intends to render the animal better suited for his service.

The tuition of a colt should commence within three months

of its birth, at which time a small bridle and surcingle may be used. Such things could at first be made of calico, the intention being merely to indicate those trappings which hereafter have to be assumed.

After a time, the above-mentioned trappings should be replaced by a small saddle and bridle, when the youngster can be led about a meadow. Kindness in breaking is all essential when confidence has to be won and timidity overcome. The hand of the master should be frequently passed over the body, and carried down the limbs; and although regular grooming is not as yet necessary, yet the mane and tail can be combed, the ears pulled, the attendant taking care to praise the foal when he seems to submit with pleasure to such operations.

THE FIRST HARNESS PUT ON A FOAL.

When the weaning has been accomplished, the colt should be well nourished in a loose box—not turned out and neglected until the determined time of breaking comes round,—and its

22 *

previous lessons must be enforced with greater emphasis as age progresses, and should be continued until the colt has passed his second year, when a dumb jockey can be mounted on his back, upon the extended points of which an old hat and cloth may be affixed. These objects will at first excite terror, but fear not being justified, confidence will return.

RECEIVING THE FIRST LESSON.

A sack stuffed with straw, and moulded somewhat into the shape of a man, should then be placed over the dumb jockey. Little stirrups and a pair of representative legs should hang on either side, while, to complete the whole, reins may be fastened to the bit; a portion of these last being formed of india-rubber, for not a few mouths are permanently destroyed by the unyielding tug of the heavy-fisted Breaker.

All these liberties being permitted, if the instruction has been properly communicated, the pupil will have been rather pleasurably excited than permanently alarmed by the varied progress of its tuition. Such lessons, however, should be daily given until the colt has attained its third year.

At three years and a half old a very diminutive lad may be put into the colt's saddle; but as boys are too apt to spoil the mouth by hanging back and holding on by the reins, the india-rubber had better be continued, and the jockey instructed not to interfere with the bridle, save when his so doing is neces-sary to guide the animal. Then the teaching of different paces may begin, the quadruped being always instructed in company with a perfectly trained old horse. All feeble intellects are apt at imitation, and a colt will readily learn from example what coercion will fail to impart.

By the fourth year the animal may be placed between the shafts of a very light gig, should its form indicate the creature not to be adapted for the saddle: at first it must be walked about a meadow. When the sound of the wheels is not listened for with evidence of fear, the pace may be quickened. Sub-sequently, a boy may get into the vehicle, while the man remains at the colt's head. Succeeding this, the course should be directed by the driver; ultimately, after a man has for some weeks assumed the office of director, the vehicle may be taken upon the road.

Most harness horses are very imperfectly broken. The edu-cation is too hurried, and seems to be considered as perfected whenever the animal will merely take to the collar. The con-sequence is, there are more bad harness horses to be met with in London than creatures of any other description. Some have all spirit lashed or jaded out of them; these have become "slugs," or the poor wretches are almost dead to command, and insensible to the goad. Others are rendered incurable kickers by the treatment to which they have been subjected. A third class are ruined by the unscrupulous use of the reins; and some of these will take long journeys, all the time holding

the bit between the teeth.   A fourth set are rendered cripples
by the unfeeling employment of the bearing-rein, which disables the organs of respiration, and renders the lightest draught
a terrible burden, by throwing the work upon the muscles of
the limbs, while it compels these agents to contract at a fearful
disadvantage.

Those who delight in a lofty crest may accomplish more by
attention to the health and diet than by the absence of humanity.
The strongest bearing-rein and the sharpest bit cannot exalt
the head of a spiritless horse.   Clover, tares, beans, and peas,
by promoting the strength and lending tone to the muscular
system, will do more to raise the neck and promote gaiety of
spirit than the harness-maker can accomplish.   Bearing-reins
are disgraceful cruelties, and do no more than expose the
moral condition or the pecuniary meanness of those parties
who employ them.

In corroboration of the importance of the neck as an aid
to motion, the reader must pardon the writer if he refers to a
well-marked circumstance, which has hitherto escaped observation.   A horse with a thin or narrow neck, measuring from
the crest to the windpipe, should always be avoided.   It
denotes bodily weakness, and testifies to an absence of spirit.
The cervical region always first exhibits the token of approaching emaciation.   If the reader will, hereafter, test the
remark by observation, he will find all poor, exhausted
animals, which carry the head as though its weight was
oppressive, invariably have the neck much impoverished
and altogether attenuated.

In short, a mere catalogue of the evils engendered by the
injudicious breaking of draught horses would occupy more
space than we can command.   For this reason, the driver
of a young animal should never be entrusted with reins
made entirely of leather; a part of the length should be
composed of india-rubber.   Neither should he be permitted

to flourish a whip. All severity is but an indulgence of the controller's temper; it is unnecessary with a life which is eager to learn and is anxious to obey. The sound of the voice, or the gentlest indication, should be sufficient to excite the ability of such a pupil. No one can doubt this, who has beheld its activity of ear, whenever the horse is addressed.

After the foregoing fashion, the education may be perfected, without allowing any professing brute, under the name of a "horse-breaker," to spoil the temper and to lay the seeds of future disease, by ill-treatment of a few weeks' duration. Some years ago, we remember meeting a man, who must have weighed more than fourteen stone, seated on a side-saddle, and having a horse-rug dangling about his heels. He was supposed to be "breaking in" a colt, rising three, for a lady equestrian. His employer must have been excessively developed, or, her representative could only spoil the creature, which was, ostensibly, preparing to receive a lighter burden and a more delicate hand. An accident was, thus, almost rendered certain, whenever the oppressed quadruped should be relinquished to its future mistress.

The matters which have already been pointed out being attended to, and the force having been increased with the growing strength of the colt, the creature, after its fifth year (if intended for the higher purposes of the saddle), should be taught to leap. To place a rider on an animal's back, and then to expect a bar to be cleared, is very like loading a young lady with a sack of flour, as preparatory to a dancing lesson being received. This folly is, however, universally practised; so is that of teaching the paces, when the quadruped's attention is probably engrossed by the burden which the spine has to sustain.

Leaping is best taught by turning the horse into a small paddock, having a low hedge or hurdle fence across its centre. A rider should, in sight of the animal, take an old horse over this several times. The groom, who brings the corn at the meal hour, then goes to that side where the animal is not, and calls, shaking up the provender all the time his voice sounds. The boundary will soon be cleared. When half the quantity is eaten, the man should proceed to the opposite compartment and call again. If this is done every time the young horse is fed, the fence may be gradually heightened; after six months of such tuition, a light rider may be safely placed upon the back.

RISING TO THE LEAP.

Instruction, thus imparted, neither strains the structures nor tries the temper. The habit is acquired without those risks which necessarily attend a novel performance, while a burden oppresses the strength, and a whip or spur distracts the attention. The body is not disabled by the imposition of a heavy load, before its powers are taxed to the uttermost. The quadruped has all its capabilities unfettered, and, in such

a state, leaping speedily becomes as easy of performance as
any other motion.

CLEARING THE FENCE.

Irish horses, all being excellent jumpers, are much esteemed
in England. In Ireland, however, the fields are of small
dimensions, and gates leading to them are uncommon. It is
not unusual for a quadruped to be obliged to clear numerous
walls, before a certain pasture can be gained. Thus, to leap
is rendered a prominent necessity of equine existence: for the
steed must either jump or starve. By such a condition of
their residence is the Irish breed made conspicuous for that
activity which especially excites the admiration of English-
men. Hunting, moreover, is a favourite pastime with the
natives of the sister isle; therefore, while most Irish horses
become admirable English hunters, the best of the English
breed would be sadly thrown out by a short run in the

adjacent kingdom. There can be, however, no reason, why an English colt, if properly trained, should not become as fine a performer as the most expert or celebrated of those animals which are generally supposed to be born "fencers."

The seventh year should witness the horse taken into the active service of its master. Too early work, certainly, cripples the majority of animals; but there is not a circumstance of the many, rebuked in these pages, which does not aid, powerfully, in producing that miserable effect. All the customs about the equine race seem to be antiquated and injurious. An animal is taken up, is cast, is operated upon, is shod, is broken, and is sold, often in the course of a few weeks. What a change has to be submitted to! Every incident of life is altered: the creature is suddenly called upon to endure a new existence. Is it a matter for surprise, that nature occasionally rebels against so wholesale an innovation? Is it not a proof of the sweetness of the disposition which graces the equine race, that the majority can yield themselves up to the barbarity of such a terrible mutation?

The animal being educated, according to the foregoing description; not being forced to strain its thews and to distort its limbs, before the frame has fairly been perfected,—but being gradually brought to the mark of its requirements, and also permitted time to comprehend, before it is lashed to perform,—being allowed the benefits of practice prior to being expected to exhibit its accomplishments,—being simply treated after a manner that every grade of reason must recognize as just,—would come forth in the full possession of all its natural powers, and would distance the swarm of equine babies, which now disgrace the thoroughfares, encumber the field, and ruin the racecourse. It would be fitted to carry a man in any manly sport; and it would be able, not being distorted by bodily pains, to sympathize in the pleasures of its rider, and to share the amusement in which he delighted.

One peculiarity, illustrative of the present mode of preparing quadrupeds for exertion, is to be witnessed in most
hunting-fields. The young gentleman who pays hundreds,
perhaps, for his "mount," and whose horse has been long
under the trainer's care, is usually "*nowhere*" at the death,
although he is at liberty to choose his way and to regulate his
pace according to his pleasure. Whereas, the huntsman, on
a screw, is generally foremost in the chase.

This seeming inconsistency evidently favours those notions
which have been promulgated. The wealthy scion of aris-
tocracy usually sits upon the young beauty, while the huntsman generally bestrides the aged animal. The older steed
may be of little worth, and its blemishes may be numerous;
but it has not been exhausted under a pretence of fitting it to
endure; it has been hacked or ridden through the months
when the younger quadruped was imprisoned in absolute
idleness. The cheaper horse has been in constant requisition
to exercise the dogs, &c., and therefore its health has been
better preserved than is that of the gentleman's steed, which
is either new to the sport, or has recently been taken from the
supposed enjoyment of a *summer's* rest.

Training of hunters and of racers, as at present conducted,
is neither a strengthening nor a refreshing process. The animal that has recently been relinquished by the trainer, instead
of being able to endure extra exertion, is generally debilitated
by those measures which were designed to produce a contrary
effect. In the first place, three doses of physic, which are
given under a belief of their tonic efficacy, are quite sufficient
to disable any creature that, like the horse, is possessed of a
very large and a very long digestive track, or which Nature,
as a protection, had rendered almost safe from the purgative
operation of medicinal agents. Before the bowels of the horse
can be loosened, the primary effects of poisoning must be
established. Aloes is the favourite purgative of the stable;

but so nearly related are the quantity which relaxes and the amount which kills, that, probably, aloes have poisoned more horses than all other drugs in the Pharmacopœia.

The reader to whom such a subject is a novelty may inquire, what the intestines have to do with the muscular action? Supposing such a question possible, we reply, that although the animal body is made up of numerous parts, and composed of various organs, nevertheless, the whole is so united that no part or structure can be diseased, but the whole is affected. The intestinal track is lined with mucous membrane. When this surface is involved, prostration or debility ensues. Cold and sore throat are ready instances of this result; for both are consequent upon small portions of inflamed mucous membrane. Imagine, then, the utter prostration which must ensue upon the morbid excitement of so large a mucous surface as that which covers the digestive canal of a horse. Yet, the trainer thrice induces this consequence, under an ignorant conviction that, by so doing, he confers upon the sufferer extraordinary nervous energy !

Purging is, however, only slightly more weakening than sweating. Perspiration acts differently on different specimens of the same species. One person is nearly always bathed in moisture; another invariably presents a dry skin. This shall hardly be moved without the surface of his body being loaded with copious drops of fluid exudation; that will endure the utmost exertion, grow heated at any employment, but will not sensibly lose a particle by transmission. The trainer, nevertheless, treats all animals alike. He gallops every quadruped submitted to his care, as though the consequence was invariably beneficial. In vain does one horse break down; another refuse its corn, and a third exhibit swollen legs or crippled feet, while a fourth shall be only rendered more lively by the process which disabled its fellows. To sweat is a part of the trainer's system, and all the creatures which he is to train must therefore be violently sweated.

With racers, to these modes of debilitating is united a third, —excessive labour. The horse is tried at its topmost speed. These trials are frequent. Although it is a common saying that a horse may be trained until it cannot move, still the practice is continued. The pace is quite as severe as it is in a public race; the weight is usually pretty much the same. It is well known that these trials are often run in less time than the contest for which they are thought to be only a preparation. Notwithstanding the repeated disappointment and the frequent injury induced, such prejudicial experiments are continued, though not in every sphere of training. Men train as prize fighters; but they do not, before entering the ring, engage in numerous pitched battles. There is, assuredly, something wrong when the same law is stringent in one case, but is inoperative in another, although both instances are supposed to be governed by the similar regulations. Some trainers are guilty of the foolish practice of starving horses just before a race, i. e. when they are most in need of support.

When extra energy is imperative, the trainer, by his conduct, pursues the measure best calculated to destroy all inclination for exertion. The plea urged in defence of such folly is, that a loaded stomach oppresses the breathing. This is true enough; but the evils which result from gluttony do not establish that good only can ensue upon starvation. Let the trainer experiment upon himself, and decide whether a light meal or no meal at all is the better preparation for an extraordinary performance. Many trainers assert that a full stomach rests upon the diaphragm, and thereby is detrimental to the respiration. This is a mistake. The digestive sac is pendent beneath the respiratory agent; a fact which an inspection of the annexed engraving will amply illustrate.

On the course weakness has lost many a race. Why should

such a system be pursued longer? Why are famishing animals,
when prostrated by the want of food, enervated by hunger

DIAGRAM, TO ILLUSTRATE THE DIFFERENT POSITIONS OF THE STOMACH AND OF THE DIAPHRAGM.
1. The lungs. 2. The stomach. 3. The Intestines. 4. The diaphragm. 5. The bladder.

and thirst, only considered qualified to exhibit fleetness? Is
not the idea, when plainly stated, a self-evident fallacy?

Again, wherefore should the hunter, when the season is
over, be shut up in a loose box, be doomed to a six months'
life of inactivity? It would thrive better were it gently
hacked by a considerate proprietor. Taken out occasionally,
and quietly ridden down the shady green lanes of the neigh-
bourhood. Never bustled, but sometimes breathed over an
even piece of turf. Ridden always for pleasure, but never
saddled when business is to be transacted. Such a life might
not allow the groom so much leisure; but it would materially
lessen his labours when the hunting season approached. The
animal would need but little "*conditioning*." Improper
sustenance would not have induced dropsy; nor would the
joints have stiffened by a long period of enforced inactivity.

A better system than that of keeping hunters in loose boxes
during the summer months is to turn them out to grass during

such period; not in a too luxuriant pasture, but in grass sufficient, together with two feeds of corn each daily, to keep them
in "good case." At grass the horse is able to roam at large,
to take exercise at pleasure, to breathe the fresh air, and to
enjoy an entire change of scene. These produce an invigorating effect on the whole system, and the cool grass exercises
a sedative influence on inflamed and injured feet.

In concluding, let us warn all horse-keepers against the
employment of incompetent persons as physicians for their
horses, who pretend to cure all diseases; and for the same
reason, never permit grooms to administer medicines unless
under the advice of a duly qualified veterinary surgeon.

Aloes and nitre are the chief perils of the stable. More
horses have diseases of the kidneys through the abuse of nitre
than would be affected if left entirely to Nature. As to aloes,
the poisonous and perilous nature of that drug has already
been dwelt upon; the pitiable infatuation with which grooms
regard it constitutes one of the heaviest and needless extrava-
gances of every hunting establishment.

The reader, as he reviews the topics submitted to his judgment, is probably surprised to perceive how little mystery
legitimately appertains to the horse, but how much its requirements accord with dictates of common sense.

Yet, is it not surprising that society at large regards "horse
knowledge" as a mysterious attainment, to be gained only by
a long course of actual experience? Such a prejudice is without the slightest foundation. Antiquated customs and exploded
notions are common enough in every mews. Filth is, in such
places, supposed to be endowed with strange medicinal attributes; and cruelty is patronized, as though the perfection of
wisdom lay in the total absence of humanity. The horse, as
at present treated, is the victim of ignorance, and is exposed
to every abuse. Nature and her dictates are disregarded. The
animal is tortured, till it submits to abhorrent brutalities. Its

instincts, as a created being, are not respected; neither are its necessities, as a living creature, ever considered. Its welfare is secondary to the convenience of the master, and its custody is transferred by him to his groom.

"*Common sense*," however, demands these things should be amended. And the main purpose with which the foregoing pages have been indited was with a hope, through a plain statement of facts and an appeal to the reason of the public, of awakening those entrusted with authority over the equine species to the errors attendant upon the present system of stable management.

INDEX.

	PAGE
ACTION of shoulder-blade during progression	294
——, high, admired by ladies	295
Aftermeath, or rowen	164
Age, an important consideration	2
Air-passages can alter their dimensions	10
Aloes, their effect on horses	39, 41
——, the purgative of the stable	347
Ambulatory	240
Anatomical considerations	1
Arabian mode of fastening horse's shoe	81
Aretenoids	13
Ass belongs to the equine race	34
——, cruelties practised on	34

B

Backbone of the horse	25
Balling-irons, common form of	42
——, improved form of	44
——, Professor Varnell's invention	44
Balls, mode of administering	45
Beans, different kinds of	178, 180
Bearing-reins, cruelty of	295, 342
Bishoped teeth	125
Bishoping described	126
Bleeding	63—74
——, mode of	70
Blistering and firing	56—63
Blisters, how applied	61
Blood, effect on, of exposure to atmosphere	9
Blood-can, form of	68
Blood-stick, form of	70
Body, the, anatomically considered	24
Boiler-house, situation of	245
Bolter	284
Bones of fore limb not self-sustaining	29
—— of spine	24
Bran mashes, how prepared	39
Breaker	140, 326

	PAGE
Breaking and training, their errors and their results	336
Breeding, its inconsistencies and its disappointments	321
Brood mare	326
——, on choosing	327
——, treatment of while foaling	328
Brushing	111
Bull-necked horse	274

C

Calkins	100
Camel, eye of	16
Cart horse	288
Cartilago nictitans	22
Caustics, dangerous as internal medicines	48
Channel, narrows with age	129
——, desired width of	286
Chink of the back	230
——, treatment for	231
Cistern, situation of	245
Cleansing horses	256
Clicking	115
——, shoe meant to prevent	116
Clip-shoe	110
Clipping and singeing	253
Clover hay, first crop of	164
——, second crop of	165
Conjunctiva	21
Corpora nigra present in the horse and camel	17
Cow-bellied, or pot-bellied horse	292
Cribbing, induced by confinement and indigestion	187
Cutting	111
——, remedies for	113

D

Deglutition	14
Dentine	149
Down on the hip	199
Draught horses, hints on purchasing	316

23

PAGE

Drinks, or draughts, as horse
　physio　.　.　.　.　49
——, danger of administering　.　50
——, for heaving at the flanks　.　330
——, process enabling horse to
　drink detailed　.　.　.　51
——, modes of administering　.　53
Dumb jockey　.　.　.　340

E

Ear of horse　.　.　.　281
Enamel　.　.　.　.　149
Epiglottis　.　.　.　13
Ewe-necked horse　.　.　275
Eye of horse　.　.　.　282
——, darkness does not incapaci-
　tate　.　.　.　18
——, endowed with telescopic
　power　.　.　.　24
——, evidence of original habitat　20
——, operation of muscles on　.　19
——, pupil of never circular　.　17

F

Field peas　.　.　.　.　178
Fleam, form of　.　.　.　69
Foaling　.　.　.　328
Food　.　.　.　.　160
——, arrangement of course of　.　13
Foot, wall of　.　.　.　79
——, paring and rasping　.　.　123
Fore-limb, how united to trunk　28
——, bones of not self-sustaining　30
Frog-pressure shoe　.　.　109
Frosting, or roughing　.　.　118
Fullered shoe　.　.　.　102

G

Gland, lachrymal　.　.　21
Grass suggests the food of the
　horse　.　.　.　.　3
Grooms, hints to　.　.　248
Gruel is proper for an exhausted
　horse　.　.　.　109
Gutters in stables　.　.　235

H

Hand over hand pace is bad　.　318
Haunch, capability for work
　chiefly dependent upon　.　311
Hay, aftermeath.　.　.　164
——, clover, first crop　.　164
——, clover, second crop　.　165

PAGE

Hay, heated, or mowburnt　.　166
——, from legumens　.　.　180
——, lowland　.　.　163
——, musty　.　.　167
——, new, is objectionable　.　163
——, promotes fat　.　.　180
——, upland　.　.　162
——, weather-beaten　.　.　167
Head　.　.　.　12
——, position of when feeding　4, 6, 7
Heated or mowburnt hay　.　166
Herring-gutted horse　.　.　292
Hollow-backed horse　.　.　271
Hoof, high or stubborn　.　306
——, horn of described　.　79
——, upright　.　.　307
——, odd　.　.　303
Horse balls　.　.　38
—— beans　.　.　178
—— dealers　.　.　131
Horses, age of, an important con-
　sideration　.　.　2
——, backbone of　.　.　25
——, body of, anatomically con-
　sidered　.　.　1
——, bull-necked　.　.　274
——, cart　.　.　288
——, cleansing　.　.　256
——, cow-bellied　.　.　292
——, draught, hints on purchas-
　ing　.　.　.　316
——, ewe-necked　.　.　275
——, fondness for malt liquor　.　258
——, frosting or roughing　.　118
——, herring-gutted　.　.　292
——, hollow-backed　.　.　271
——, Irish, good fencers　.　345
—— naturally gregarious　.　345
——, neck of, an aid to motion　.　342
——, origin of　.　.　16
——, paring and rasping foot of　123
Horses, precautions to observe
　on purchasing　.　.　284
——, progression of　.　.　85
——, roach-backed　.　.　272
——, stomach, capacity of　.　3
——, teeth of　.　.　125
——, well-bred　.　.　278
Humerus　.　.　29
Hunter, muscular loins impera-
　tive in　.　.　.　270

I

Incapacitating vices, the results
　of injury or disease　.　.　224

Incisors at different periods of
   horse's existence . . 133—146
Intestines, extent of . . . 3
Irish horses good fencers . . 345
Itching and scratching . . 190

J

Jaw of horse . . 129—130
Jibbing 233
Jointed shoe . . . 108

K

Kidney dropping . . . 225
——, treatment of . . 227
Kiln-dried oat . . . 176
Konigsberg oat . . . 171

L

Lachrymal gland . . . 21
Laminæ 80
Lampas 149
Laryngeal affections, annoyances
   of stables . . . 5
Larynx, composition of . . 10
Leaping, instructing young horse
   in 343
Legumens used as hay . . 180
Legs of horse can hardly prove
   too short . . . 293
Liniment to be used after blis-
   tering . . . 62
—— for rick of back . . 231
Loins cannot be too large . . 269
——, spines of, and of the sacrum
   point different ways . 32
Loose boxes are preferable to
   stalls 235
Lowland hay . . . 163
Lumbar 25
Lunging 337
Lungs, capacity of, illustrated . 2

M

Malt liquor, horses grow fond of 258
Molar teeth or grinders . . 146
Motor agents . . . 20
Mouths of horses . . . 283
Mowburnt hay . . . 166
Muscles, operation of, on the eye 19
Musty hay 167

N

Navicular disease . 85, 94, 206
Neck, an aid to motion . . 342
Nibbling woodwork, induced by
   enforced idleness . . 186
Nippers, a misnomer . . 159
Nitrogen 8
Nostrils . . . 283, 285
——, false . . . 9
——, size of . . . 9

O

Oat, characteristics of sound . 176
——, different species of . 170, 178
——, hair calculi . . 176
——, sulphur in . . 176
Odd hoofs . . . 303
Os hyoides . . . 287
—— innominatum . . 268
Ox, his mode of feeding . . 4
Oxygen 8

P

Paring and rasping horse's foot . 123
Pastern bones, where they repose 31
—— are intended to endow the
   tread with elasticity . 301
—— are regulated by the flexor
   tendons . . . 302
Pendulous palate . . . 14
Perspiration implies cuticular
   activity . . . 255
——, insensible, is attended with
   danger . . . 256
Petersburg oat . . . 171
Pharynx 13
Physic, administration of . . 38
Points, their relative importance,
   and where to look for their development . . . 268
Progression of the horse . . 85
Pupil of horse's eye never circular 17

Q

Quarters, hind, the seat of propulsion 32

R

Racer, muscular loins imperative
   in 270
Rasping horse's foot . . 123
Respiration 2

	PAGE
Ribs of horse	27
Rick or chink in the back	230
———, treatment for	231
Roach-backed horse	272
Roman nose	278
Roughing	118
Rowen hay	104

S

Sacral bones, how fixed	33
Sacrum, osseous junction of	25
Sand in the eye, how protected against	20
Scapula	29
Screw shoe	108
Seated shoe	107
See-sawing, or weaving	187
Shoe, Arabian	75
———, clip	110
———, French	95
———, fullered	102
———, jointed	108
———, Moorish	76
———, Persian	76
———, Portuguese	76
———, Old English	77
———, screw	108
———, seated	107
———, stamped	95
———, three-quarter	106
———, tip	103
——— to prevent cutting and clicking	115, 116
——— to enable horses to work on slippery roads	120, 121
———, wedge-heeled	109
Shoeing, its origin, uses, and varieties	75
———, English mode of	82
Skeleton	268
Sloping floors	209
Soft palate	13, 14
Spavin, signs of	318, 320
Speedy cut warrants instant rejection	318
Spine, how the bones of are united	25
Splint, injuries to, how detected	227
———, signs of	318
Stables, evils of	183
——— as they should be	235

	PAGE
Stomach, capacity of	3
Sulphur in oats	176

T

Tapidum lucidum	18
Teeth, their natural growth, and abuses to which they are subject	125
——— at advanced periods	154—159
———, bishoping	128
———, front milk	153
———, molar or grinders	146
———, wolves'	136
Thorax, best form of	288
———, framework of	27
———, narrow	293
———, depth of indispensable for speed and activity	308
Thorough-bred horse, characteristics of	280
Three-quarter shoe	106
Tips	105
Tip shoe	103
Tushes	156
Twisted suture	73
Twitch	53

U.

Undulating pavements	237
Upland hay	162

V.

Valves in veins of leg	204
Veins, arrangement of	6
———, uses of	8
Ventilation of stables	246
Vertebræ	24, 268
Vices, incapacitating, so called	224

W.

Wall of foot	79
Water-troughs preferable to pails	245
Weather-beaten hay	187
Weaving, or see-sawing	187
Wedge heeled shoe	109
Windpipe, form of	11
Wind-sucking induced by stables	187
Wolves' teeth	136

WYMAN AND SONS, PRINTERS, GREAT QUEEN STREET, LINCOLN'S-INN FIELDS, LONDON, W.C.